A PLUME BOOK

POSTCARDS FROM MARS

JIM BELL is a professor of astronomy at Cornell University. He remains the lead scientist for the Pancam color imaging system on the NASA Mars Exploration Rover missions. He lives in Lansing, New York.

Praise for *Postcards from Mars*

"Jim Bell has compiled a book of photos that puts our neighboring planet in sharp focus."
—*USA Today*

"[A] stunning portfolio of robotic snapshots that capture the red planet in all its desolate beauty."
—*Entertainment Weekly*

"The crystalline photographs can't help but impress."
—*Houston Chronicle*

"The quality and volume of photos coming back from *Spirit* and *Opportunity* far exceeds anything that has come back from Mars before."
—*American Photo*

"Postcards from Mars is a coffee-table book of fabulous photos that also includes the compelling human story behind the successful *Spirit* and *Opportunity* rovers. . . . *Postcards from Mars* is a handsome gift book, and an excellent tool to jumpstart imaginative trips to Mars and discussions of space exploration and settlement with family, friends, and coworkers. Highly recommended."
—National Space Society

"[Bell] tells the story of preparing the spacecraft for launch and of the rovers' arrival on Mars in an unconventional tone, with just the right mix of detail and personal point of view. . . . The quality of the printing is high, reproducing intense colors, dramatic shadows, and fine details."
—The Planetary Society

"Awe-inspiring. . . Exquisite detail. . . Glorious."
—*Natural History*

"This superbly produced set of photographs is a spine-tingling preview of the dramatic landscape that future humans will one day explore in person. Jim Bell has put together a beautiful, inspiring, and powerful book."
—Dr. Buzz Aldrin, one of the two astronauts aboard the *Apollo 11* lunar module, the first manned mission to land on the Moon

"*Postcards from Mars* is a masterpiece. It promises the future for all of us."
—Ray Bradbury, author of *The Martian Chronicles* and *Fahrenheit 451*

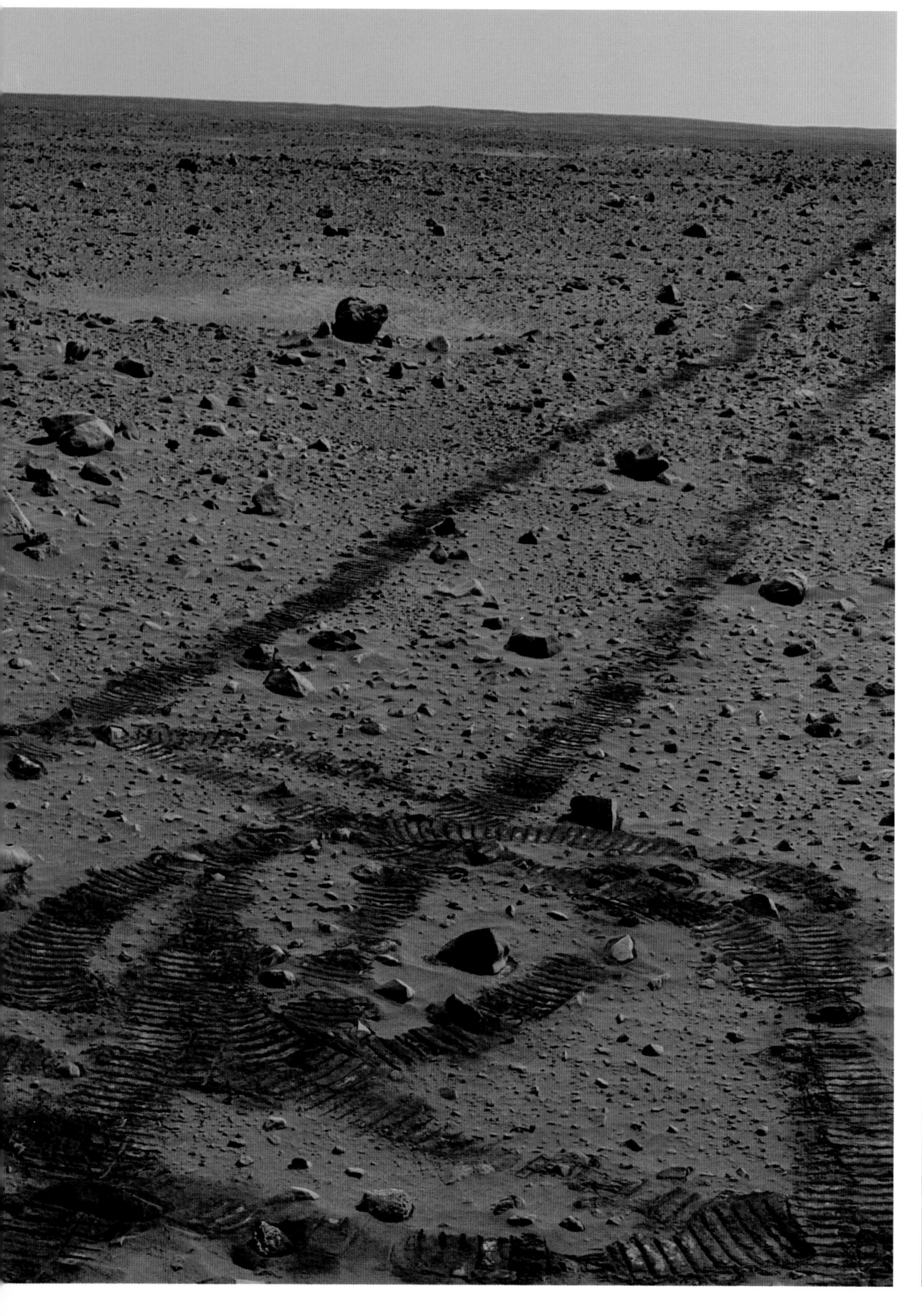

POSTCARDS FROM MARS

The First Photographer on the Red Planet

Jim Bell

A PLUME BOOK

PLUME
Published by Penguin Group
Penguin Group (USA) Inc., 375 Hudson Street, New York, New York 10014, U.S.A
Penguin Group (Canada), 90 Eglinton Avenue East, Suite 700, Toronto, Ontario, Canada M4P 2Y3 (a division of Pearson Penguin Canada Inc.)
Penguin Books Ltd., 80 Strand, London WC2R 0RL, England
Penguin Ireland, 25 St Stephen's Green, Dublin 2, Ireland (a division of Penguin Books Ltd)
Penguin Group (Australia), 250 Camberwell Road, Camberwell, Victoria 3124, Australia (a division of Pearson Australia Group Pty Ltd)
Penguin Books India Pvt Ltd., 11 Community Centre, Panchsheel Park, New Delhi – 110 017, India
Penguin Group (NZ), 67 Apollo Drive, Rosedale, North Shore 0632, New Zealand (a division of Pearson New Zealand Ltd.)
Penguin Books (South Africa) (Pty.) Ltd., 24 Sturdee Avenue, Rosebank, Johannesburg 2196, South Africa

Penguin Books Ltd., Registered Offices: 80 Strand, London WC2R 0RL, England

Published by Plume, a member of Penguin Group (USA) Inc. Previously published in a Dutton edition.

First Plume printing, November 2010

10 9 8 7 6 5 4 3 2 1

The Library of Congress has cataloged the Dutton edition as follows:
Bell, Jim 1965–
Postcards from Mars : The First Photographer on the Red Planet / Jim Bell
p. cm.
Includes bibliographical references and index.
ISBN 0-525-94985-2 (hc.)
ISBN 978-0-452-29674-9 (pbk.)
1. Mars (Planet)—Pictorial works. I. Title.
QB641.B45 2006
559.9'230222—DC22

Printed in China
Original hardcover design by Matthew Schwartz Design Studio
Art Director: Matthew Schwartz · Designers: Matthew Schwartz & Abigail Smith

DEDICATION

To my family, for their love and support during my trip(s) to Mars, especially to Maureen, for her infinite patience. And to my extended rover family, here and there, for all your hard work and camaraderie.

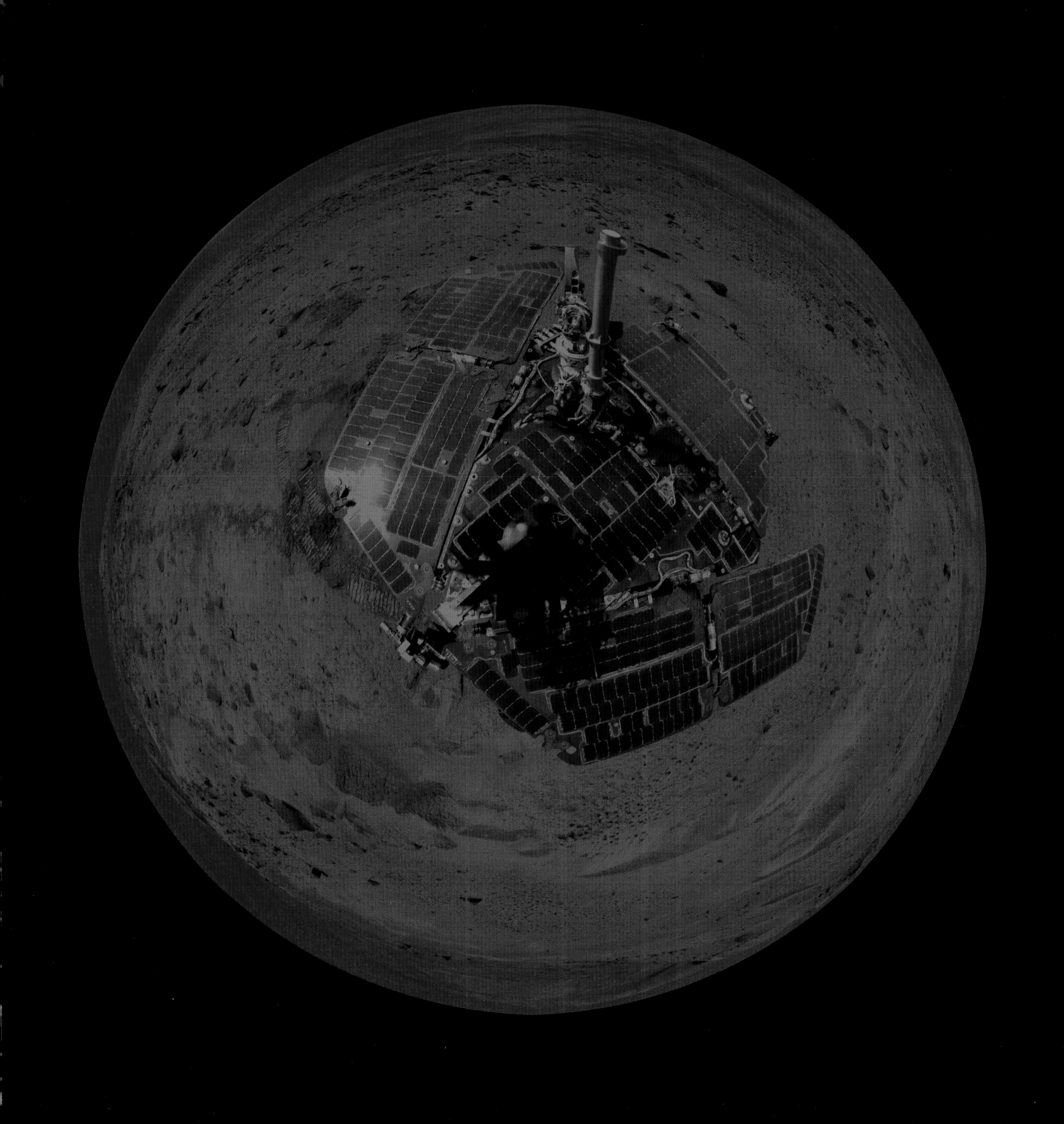

CONTENTS

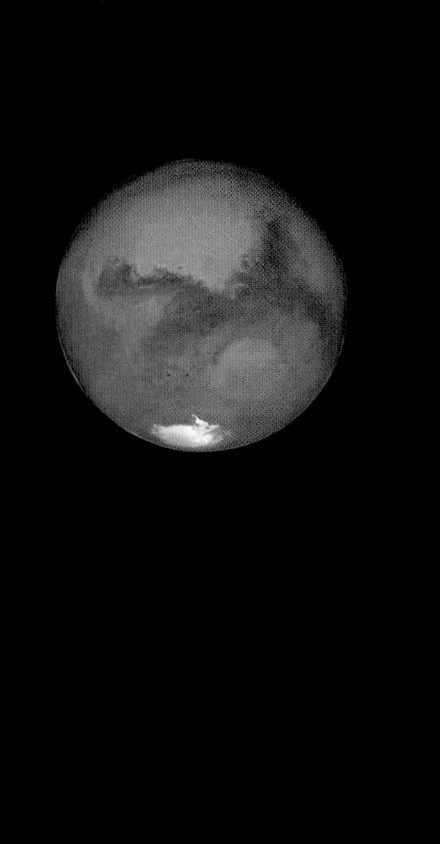

Foreword

When you see Mars through a telescope, it seems awfully far away. It is—70 million kilometers when it's close. But now, take a look at a picture taken on the surface of Mars, postcard-style. Mars is a place, hardly remote at all. If you were dressed properly, you could stand there. If your space suit were comfortable enough, you'd find walking easy, as there's less than 40 percent of the gravity we're used to. You could pick up a few souvenir rocks and maybe stop for lunch—a very quiet meal. If you really could get there, some of the pictures you would take to send home might look like these: postcards from Mars.

Open the book, and look at the martian sky. By an earthling's standards, it's an odd color, to be sure, but its otherworldly quality comes from something else. There is no hint of movement, no evidence of plants or people: no airplanes, no trees, nor buildings along the horizon. It's stark and yet quite beautiful. Somehow, though, it's familiar.

Look at the rocks. The surface of Mars is littered with them. Something must have broken them loose from the hard crust. In geology, it is said, "Every rock tells a story." With that, then, anywhere on Mars speaks volumes. It looks like there were countless impacts, innumerable rocky hammers hurtling in from interplanetary space that shaped the surface of our nearest neighbor. No movement now, but on many occasions there was high drama.

Look at the sand. Where did it all come from? On Earth, sand is created from solid rocks by the daily grind of the wind and the sea. We can observe windstorms on Mars from here on Earth. But perhaps Mars was once a wet world as well as a windy one. Water waves may have steadily ground rocks into sand. Though perhaps not from the drama of hurtling meteorites and solid rock brought from sub-freezing to above their melting point in a few milliseconds, evidence of these dramatic changes is here, nevertheless.

A geologist on Earth carries three essential tools out into the field...or the mountains, or the desert: a hammer to see what's inside the rocks, a small magnifying lens to look closely at the grains and crystals of the specimens, and a pair of boots so he or she can get around in rocky terrain. Well, our *Spirit* and *Opportunity* rovers have the same equipment. Instead of a hammer, they have small rock grinders. Instead of a hand lens, they have cameras fitted with magnifying lenses. And instead of getting mobile with boots, they use their wheels. But the idea is the same. To learn about a planet's past, we look at the rocks. You can't do it as readily from a spacecraft; you've got to get down in the dirt. That's the fundamental idea behind sending spacecraft to Mars.

Almost everyone has pondered the possibility of life on other worlds. I suppose I started exploring Mars in a scientific fashion when I was in college. My astronomy professor was Carl Sagan. Every day, it seemed, he brought in astonishing pictures from space. In the spring of 1977, the pictures were from the *Viking* landers on Mars. Professor Sagan often discussed "exobiology," the study of life elsewhere. It sounds at first like a trivial pursuit, one that is about places literally pretty far out and hardly important in our day-to-day lives. But even then, one could readily see that Mars had channels and valleys, huge wide plains; it looked like a place where water had once flowed. What if it had? What happened there to make all the water disappear? Could it happen here on Earth? Water, the essential compound for life on our world, is missing. Why? That's about as relevant as a question can get.

When studying a new field site, one of the first things a geologist does is observe the color of the rocks. Red or orange often mean rust, one form or other of oxidized iron minerals. That often leads to speculation about the rocks having once been wet. On Mars, sunlight is somewhat different from the sunlight on Earth, though. It can mislead a planetary geologist's interpretation of a rock's true color quicker than a hammer can cleave a crystal in basalt.

If you've never tried this, go outside on a sunny day. Cast a shadow on a white piece of paper (try using your fingers or a pencil). You'll observe that the shadow is gray, but it's also ever so slightly light blue. A few photographers and artists notice this phenomenon. But most of us live our lives unaware of the contribution to our view of things by our own planet's blue skies. Martian skies have their own unusual contributions. Jim Bell had an elegant idea for handling the effects of skylight on the martian surface. You'll see it among these postcards from Mars.

I got involved in the amazing adventures of *Spirit* and *Opportunity* in 1998, about twenty-one years after I took Astronomy 101 from Carl Sagan. Jim and I had never met, but while on a flight together to Ithaca, he had the flight attendant hand me his business card. He invited me to a meeting about the *Mars Exploration Rovers*. I enthusiastically accepted. A meeting about Mars? In the same conference room where I heard Professor Sagan discuss exobiology years before? Count me in.

Although he knew me as the host of a TV show that he and his kids enjoyed, I am not sure Jim knew I'm also a mechanical engineer. I used to make an honest living at the Boeing Commercial Airplane Company. So when I saw the engineering drawings for the rover camera's color-balance target, and when I found out that it used a metal post to cast a shadow to help correct for color changes caused by the martian sky, I became passionate. See, my dad had been a prisoner of war, captured from Wake Island in the Pacific and taken to Mainland China. With no electricity, he and his fellow POWs had very dark skies. He could identify perhaps forty constellations (very few modern astronomers can get past nine or ten). He also spent time trying to figure their latitude, their position on the Earth, using nothing but the shadows of sticks—things like fence posts and shovel handles. I don't know how much success he had, but he came back to the States fascinated with sundials. I was raised with the family car being pulled over so he could photograph dozens of otherwise hardly-known sundials. So, when I saw this post intended to cast a shadow on Mars, I could hardly believe it. I thought for a few moments that Jim was teasing me, pulling my leg, yanking my chain. A shadow-casting feature? Jim—everyone!—it's a sundial! We have to turn this into a sundial! And that's how the "MarsDials," carried by the rovers *Spirit* and *Opportunity*, were born. They're now the most photographed human-made objects ever sent to Mars, and it's still a thrill every time I see a new picture of one of them, covered in dust or stretching its thin shadow across the landscape in the late afternoon Sun.

Even though we didn't have to modify the rovers' photometric calibration (CAL) targets very much to turn them into MarsDials, Carl Sagan's legacy was never far from our minds while we were doing it. That's one reason why we added some special inscriptions to the MarsDials, including the only message to the future on these spacecraft. Dr. Sagan often pointed out that if we're going to toss something into the cosmic ocean, we should include something about ourselves. It's akin to a message in a bottle. You write such a message for yourself, for your own satisfaction. But there is the irresistible human hope that someone will find it someday and appreciate your effort. Perhaps a reader of this book will go to Mars one day and read our inscriptions there:

> People launched this spacecraft from Earth in our year 2003. It arrived on Mars in 2004. We built its instruments to study the martian environment and to look for signs of life. We used this post and these patterns to adjust our cameras and as a sundial to reckon the passage of time. The drawings and words represent the people of Earth. We sent this craft in peace to learn about Mars' past and to prepare for our future. To those who visit here, we wish a safe journey and the joy of discovery.

To those who visit *Spirit* and *Opportunity* through these photographs or one day in person: safe journeys and joyous discoveries.

BILL NYE
LOS ANGELES, CALIFORNIA

Preface

We have an unknown distance yet to run; an unknown river yet to explore. What falls there are, we know not; what rocks beset the channel, we know not. Ah, well! We may conjecture many things...With some eagerness, and some anxiety, and some misgiving, we enter the canyon below, and are carried along by the swift water.

John Wesley Powell
The Exploration of the Colorado River and Its Canyons, 1875

We were now about to penetrate a country...on which the foot of civilized man had never trodden; the good or evil it had in store for us was for experiment yet to determine...entertaining as I do, the most confident hope of succeeding in a voyage which had formed a darling project of mine for the last ten years, I could but esteem this moment of my departure as among the most happy of my life.

Meriwether Lewis
Journal entry; April 1805

John Wesley Powell and his motley crew of men who set out to explore the Colorado River in 1869 admitted that they had no idea what was coming. And yet, they walked, and waded, into the adventure knowing full well that they may not return—indeed, some of them didn't. As I think back on the last decade's adventure of designing, building, testing, launching, and operating on Mars two complex and hardy robotic space vehicles called *Spirit* and *Opportunity*, I can't help but wonder if we were just as naive when we started out. None of us thought we might be risking our lives, of course, but we certainly ended up risking our health, our marriages and family relationships, our friendships, our scientific careers on some level, and sometimes, it seemed, our sanity.

Our missions were not the first to land on Mars, to rove on Mars, or to acquire images of the ruddy alien landscape. The two *Viking* landers (1976–1982; 1976–1980) and the *Mars Pathfinder* lander and its *Sojourner* microrover (1997) came before us and were phenomenal technical and scientific successes. Those missions beamed back tens of thousands of images that revealed the surface of Mars, for the first time, to be a rocky, dusty, and strangely familiar place. Our achievements with *Spirit* and *Opportunity* were only possible, to paraphrase Isaac Newton, because we were already propped up on the shoulders of these first giant missions of Mars exploration. We have not been the first to see the surface of Mars, but we have had the privilege of being the first to see the places we have visited in an entirely different, and ultimately more human, way.

The difference between the views of Mars from the *Vikings* and *Pathfinder* and the views from *Spirit* and *Opportunity* is the difference between "acquiring images" and "taking photographs." Ac-

Portion of vertical projection false-color Pancam postcard shot on SOLS 652–666 on the Olympia outcrop at Erebus Crater. (NASA/JPL/CORNELL)

quiring images is a technical, science-driven, resource-limited activity. Every space mission to a new place—whether human or robotic—has to carry a camera. These cameras are the eyes that have to tell the stories of new alien worlds to the people back home who couldn't go. They also have to be able to gather the required information about a place—size, shape, distance, number of rocks in the way, etc.—to allow scientists and engineers to run the mission and to make discoveries. But it's not easy to take these pictures or to send them home. Spacecraft and instruments are complex, sometimes finicky things to operate, and the time to take pictures is often a scarce commodity. Even more scarce, usually, is the bandwidth necessary to transmit good quality pictures back home from outer space. It's like drinking from a fire hydrant with a straw. After a while, you'll quench your thirst, but only after wasting a lot of water.

Those of us taking photographs with the Mars rovers, on the other hand, have had the luxury of much more time devoted to picture taking, much more bandwidth for sending pictures back to Earth, and better resolution of our cameras compared with that of any previous Mars missions. These advantages have allowed us to not just acquire images, but to take photographs. We can be photographers—artists—while at the same time gather all of the required scientific and engineering information needed to run the missions. When I am designing a camera sequence for the Pancams—panoramic cameras—for example, I can think about the same kinds of issues that landscape photographers consider in their quest to capture the spirit and stories of the land. How can we frame this particular shot? Can we include some foreground rover parts in the image to give the view a sense of depth? What is the balance of sky and ground? Do we view the scene in natural light or with enhancing filters? And how do we interpret the view later, in the computer "darkroom" where we process the images?

I was into landscape photography when I was a kid. My parents bought me a Pentax 35 mm SLR camera, and I spent a lot of time shooting the outdoors with my friends in the high school Photography Club. I was fascinated with the interplay of light and shadow in the environment, with the way a photograph could be framed and composed, like a musical piece, to tell a story to the viewer in a certain way. I went to the library and soaked up Marcel Minnaert's book *The Nature of Light and Colour in the Open Air* and checked out picture books about nineteenth-century landscape photographers like Timothy O'Sullivan and William Henry Jackson, and twentieth-century ones like Edward Weston and the master Ansel Adams. When I figured out how to hook up my camera to my telescope, I was hooked. Space was the ultimate landscape. That's when I knew that I wanted to get into astronomy and space exploration.

Little did I realize back then, though, that I'd have the opportunity to take some of the most spectacular photographs of martian landscapes ever made. In a sense, the pictures in this book represent the culmination of the dreams of a little kid who started fumbling with filters and f-stops in rural Rhode Island thirty years ago. I look at where it led, and I'm in hog heaven. I was trained as a scientist, but I've come back to my roots in many ways and have become a space landscape photographer. Indeed, all of us involved with the rover cameras have become photographers. Even the rovers themselves are photographers, in a way. We're the first photographers on the red planet.

When Steve Squyres asked me to take charge of the rovers' Pancams, I promised to be personally responsible not only for making sure the cameras would take great photographs on Mars, but also to be responsible for the quality and color of every Pancam picture or panorama that we released to the world. You can see the best results published on these pages. Every one of the Pancam images here, and on our rover Web sites, has taken a trajectory from Mars to Earth and then through my laptop computer (sometimes to the frustration of my colleagues) before being posted or printed. With a data set of more than 100,000 Pancam pictures, this has been, well, a bit of work. But it's been a deeply rewarding task, and I'm proud of what I believe are the closest representations yet made of what it must be like to be there, standing on Mars. Of course there is an enormous amount of technical and computer work that has gone into taking the photographs and making these images, and I've been fortunate to have had a lot of help with that work from some incredibly talented and creative people on the team. However, the artistic, aesthetic—photographic—aspects of these images are my doing. The relation to reality that all art must bear is a particularly strange one for this project. It is not abstract art, but it also isn't a reality that any human has quite witnessed yet, either. I'm proud of this part of my work, too, and whatever artistic merit or sense of realism that these photos lack is my weakness, and not NASA's.

My goal in this book is to share the beauty, desolation, grandeur, and sometimes plain old alien strangeness of the Mars that has been revealed to us through the rover cameras. In editing the enormous number of photographs we've taken down to the 150 or so included here, I chose images that were representative of different phases of each rover's journey, and of major scientific successes (or, occasionally, disappointments) along the way. I've included some of the history and the stories behind the mission and the pictures as well. Some of the panoramas are huge digital image products—more than 20,000 pixels wide; while the prints are stunning, it really takes loading them onto a computer and zooming...and zooming...and zooming...to appreciate the full quality and resolution that we were able to obtain. Most of the images are presented in an approximate true color rendering, which is an attempt to simulate, as best we can, what the view would look like had you been there looking out over the landscape yourself. It's approximate because the Pancam's color filters don't respond to colors the exact same way the human eye does, and it's a rendering because sometimes it took many days to acquire all the images in some of the panoramas. During that time the lighting or dustiness of the atmosphere may have changed, but we would often smooth over these differences, especially to avoid ugly "seams" in the sky, and render the scene in a way that simulates viewing it all at once. In general, the tweaking that we've done was to get the colors right, and the result is a good estimate of what humans—properly dressed—would see if they had been there. Some images are shown in garish, Andy Warhol-esque false color, though. This is partly because there isn't any better way to display results using our infrared color filters (which are designed to emphasize specific colors or features in the rocks and soils), and partly because I think they just look better that way. For instance, you will see about thirty of the smaller photographs highlight geological phenomena such as "blueberries," which you can see on the cover. These little gems contain hematite, an indicator that there was once surface water on the planet, but which we would not see as well without the infrared filters.

Throughout *Postcards from Mars* I have followed scientific convention and given all measurements in metric. Mixing imperial style British and metric measurements has a bad reputation in space exploration these days. In the back of the book a conversion table is provided.

Postcards. That word sounds a little small to me now, given what these photos mean to me, but that is what we called them when we shot them and as they came back from Mars, and I can't call them anything else now. Leading the team that took these photographs was one of the proudest accomplishments of my life. Working on them further in the production of this book took me to a new place in my appreciation of what humanity is accomplishing in journeying to other planets. I share these images in the hope that the postcards will keep coming.

USAF
BOEING
Sverdrup
DELTA
OPPORTUNITY
MER-B
BOEING
NASA

Part 1: The First Thing to Photograph

"Jim! Jim! It's not just a 'calibration target,' it's a sundial!" That was how the MarsDial was born. Bill Nye's eyes were bugging out of his head. Bill, Cornell class of '77, was in town to participate in the dedication of the Sagan Planetwalk, a 1:5,000,000,000 scale model of the solar system spanning the city of Ithaca. I had just been at NASA's Jet Propulsion Laboratory (JPL) in Pasadena, California, and had camera and calibration target designs still dancing in my head. When you are photographing another world it's important to have some earthly reference.

Back in the 1970s, when the *Viking* lander cameras were being designed, the scientists and engineers involved came up with the idea of carrying along a sort of calibration plate containing painted grayscale and color patches of known brightness and hue. No one had ever taken pictures from a camera on the surface of Mars before. Everyone knew that the surface and sky would probably be reddish based on telescopic and early orbiter camera views, but no one knew exactly what shade of red it would be, or if there would be rocks and soils of other colors there as well. Getting the colors right with space cameras is a tricky business. While you do the best you can with whatever tests you can do before launch, the *Viking* team realized that if they carried along an object for which they knew the colors ahead of time, then all they had to do was take a picture of that object along with the rest of the scenery. If they got the colors right on the "known" calibration plate, then they'd have the colors right on the "unknown" rocks, soils, and sky. It was a pretty clever idea, but unfortunately it backfired initially on the team: When the first pictures came down, they didn't include the calibration plate and so the colors had to be estimated based on the prelaunch tests only.

The result was a set of rapidly generated initial images that were showcased on TV to a world hungry for the first images ever from the surface of Mars, and they showed a bluish sky. Subsequent images of the calibration plate, which allowed a more proper processing of the colors in the images, revealed that the sky was actually more of a salmon-pink hue. The initial blue sky pictures were replaced by their correct reddish-sky versions, and subsequent images were processed correctly. To this day, though, conspiracy theorists continue to have a field day about how NASA has covered up the "real" blue color of the martian sky. Sigh...

Opportunity began her journey to Mars on July 7, 2003, in a spectacular nighttime launch. (NASA/JPL/KSC)

FOLLOWING SPREAD
Pausing for a photo during *Spirit*'s prelaunch checkout. (NASA/JPL/KSC)

LIFT HERE

MER1

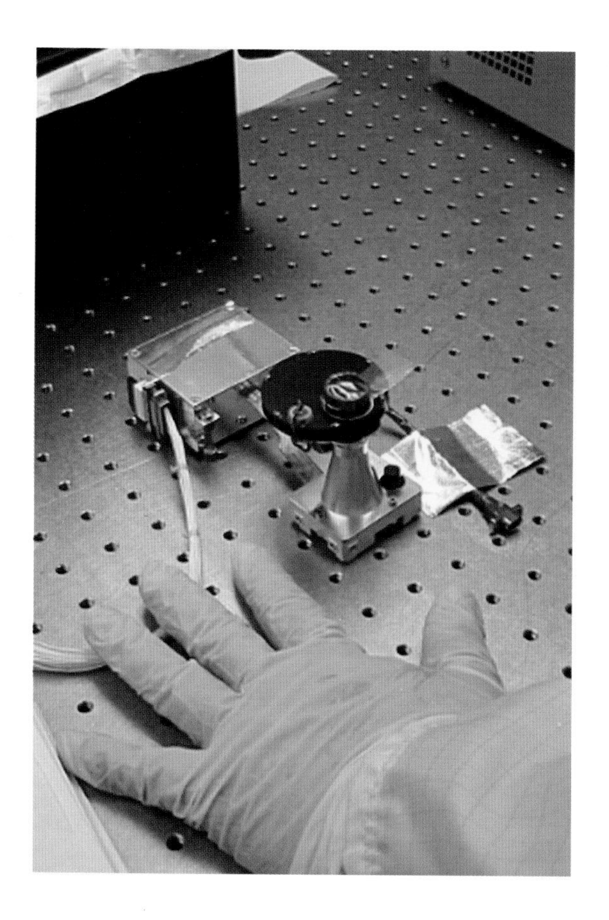

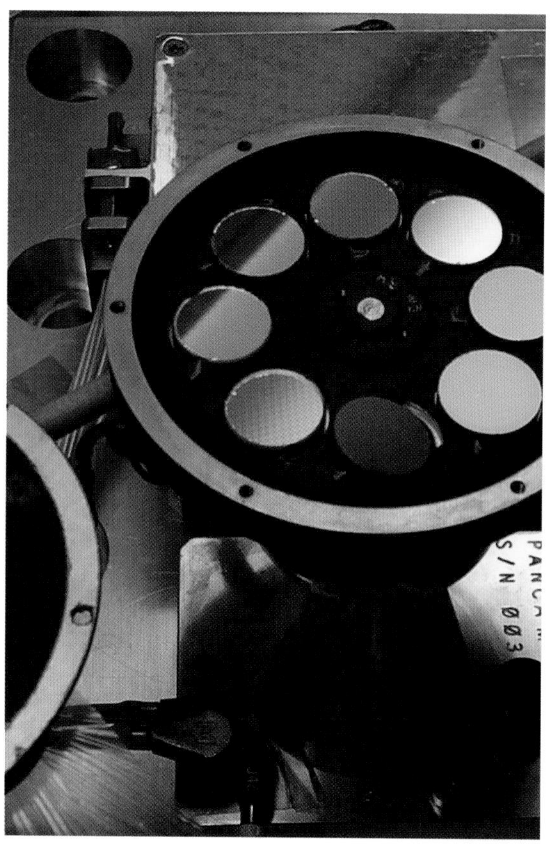

The first Pancam, newly born in the assembly and test lab at JPL; June 18, 2002. (JIM BELL)

Inside Pancam's filter wheel, showing a filter that was dislodged during initial vibrational testing. Each color filter is about 1 cm in diameter. (DAVE THIESSEN AND DARRYL DAY/JPL)

Spirit Pancam pre-launch test image of the MarsDial, a Frappucino bottle, and a series of rock and calibration samples assembled by teammate Dick Morris from the Johnson Space Center. (NASA/JPL/JSC/CORNELL)

Ten years later, when scientists and engineers were designing the cameras for the *Mars Pathfinder* lander, they decided on a similar strategy, this time designing two small calibration targets with grayscale rings and colored pieces with properties similar to the *Viking* lander calibration plates. They modified the design, though, based on lessons learned from the *Viking* experience, and added an important new feature: a central post in the middle that would cast a shadow across the target. Within the shadow, the only light would be from scattered light from the sky, so images of the sunlit and shadowed part of the target would allow the sky's color to be measured and removed from the color of the surface rocks and soils.

The *Pathfinder* calibration target design was elegant. I was deeply involved in that mission and gained a lot of experience using the target to get the colors right in the *Pathfinder* images. So when we started designing the Pancams for the *Spirit* and *Opportunity* rovers, the *Pathfinder* targets were the place to start in designing our new rover targets. We tweaked a few things from the *Pathfinder* version, but kept the basic rings, colors, and shadow post design. It would be a simple, functional, no-moving-parts device that would help provide important but esoteric data needed to calibrate the Pancam images. It didn't show up on the "fever charts" of most of the rover engineers and managers. Mostly they only cared about its weight (about 60 g) and volume (about 8 x 8 x 6 cm). I was just happy that we could find a small space to mount it on the rover deck.

I had invited Bill Nye to sit in on a meeting. I was excited about the calibration target, despite its mundane nature. I figured that Bill would be a kindred soul in appreciating the simplicity of what would be such a critical part of the camera system. I explained that it was a flat planar surface with some colored designs and a post that casts a shadow.

Bill, as most people who meet him find out soon enough, is crazy about sundials. It's an affliction that he inherited from his father, who wrote a book about them following an obsession that developed while he was a prisoner of war during World War II. I love sundials, too. It's hard to think of a simpler scientific instrument: Get a stick and use it to cast a shadow and then you've got a clock and a whole bunch of other information about the planet you happen to be on. The history of sundials reaches back thousands of years. The Greek astronomer Eratosthenes used a pair of simple sundials to estimate—with astounding accuracy—the size of the Earth more than 2,000 years ago. According to Bill, the flat planar part of our calibration target was the dial's "face" and the shadow post was actually called a "gnomon" in sundial lingo. If it was to be a real sundial, however, it would need some embellishments. Specifically, like sundials everywhere, we'd have to endow it with the date and location of its first use and some kind of motto. Also, if we were really going to be true to tradition, it would have to be adorned with other distinctive characteristics that aficionados refer to as the dial's "furniture."

We enlisted the help of some other sundial lovers we knew, including scientists, artists, and engineers who shared our desire to turn this esoteric little calibration object into an interplanetary objet d'art. Steve Squyres, the leader of NASA's whole *Spirit* and *Opportunity* expedition to Mars, and I wanted to make sure we didn't compromise the scientific integrity and calibration function of the target; after all, these were taxpayer dollars. We could be creative with the design, but in the end we had to be able to calibrate the cameras with this thing. The details of the gnomon and other furniture were worked out by Bill; astronomer Woody Sullivan, a gnomonocist extraordinaire from the University of Washington in Seattle; Larry Stark, a machinist and designer working with Woody at the University of Washington; and Lou Friedman, an engineer and one of the founders of The Planetary Society. Additionally, artistic inspiration was provided by Tyler Nordgren, a former Cornell astronomy graduate student with a talent for drawing and a good eye for aesthetics, and space artist Jon Lom-

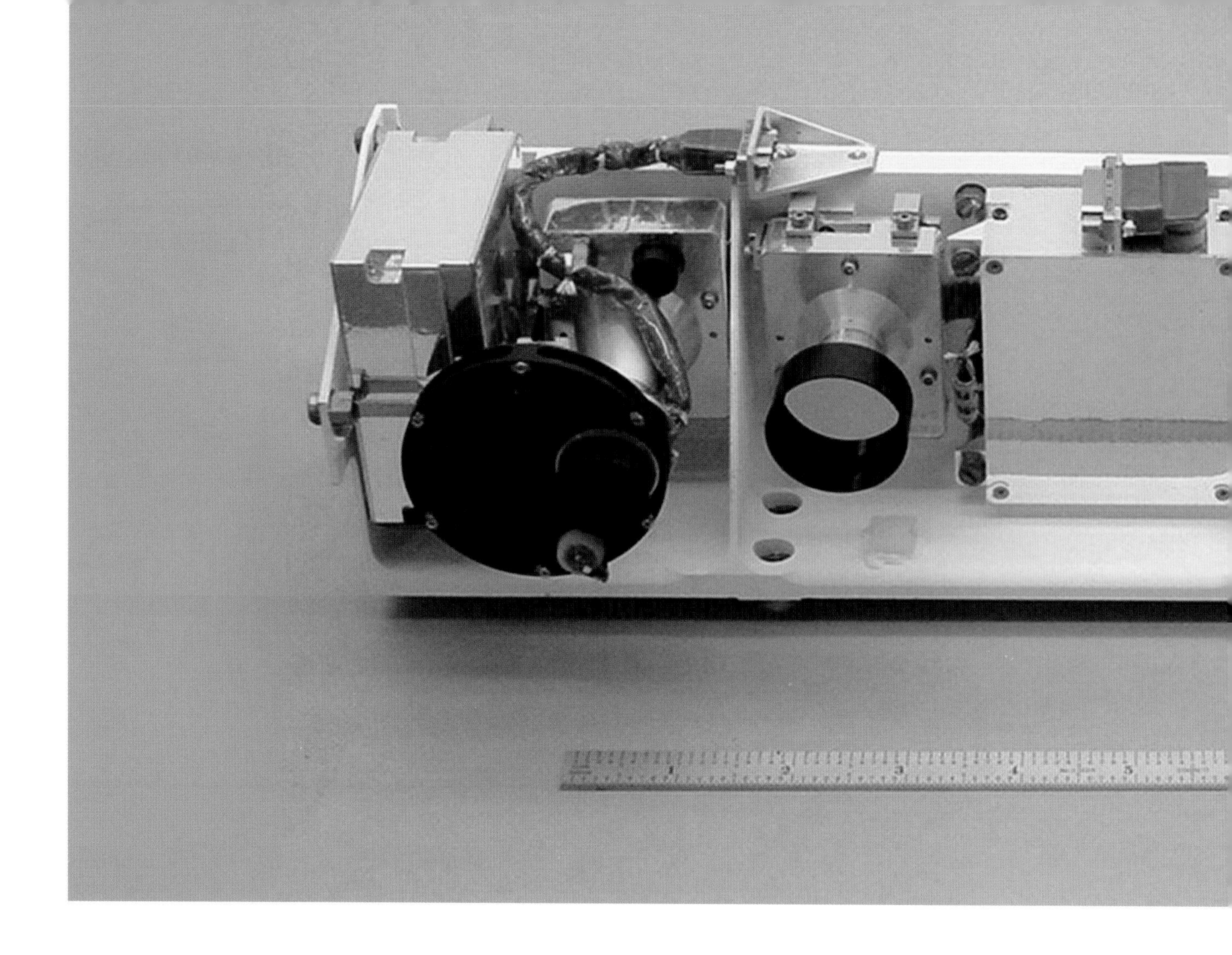

berg, who had played important roles in designing the *Voyager* interstellar record, the art for Carl Sagan's book *Cosmos* and its television series, and many other space-related art projects.

We settled on furniture that performed calibration functions well but also had some flair. The grayscale rings were resized to represent the elliptical orbits of the Earth and Mars, and small blue and red dots were positioned along each orbit to represent the two planets. These may confuse future archeologists and historians, though, because the planets are depicted according to their positions for a 2002 landing of an earlier, eventually cancelled precursor version of our mission. This was one aspect of the dial's design that we were unable to change when that mission was transformed to a 2003 twin rover launch. We added the year of the dial's first use (2004) and a unique and fitting motto to the face: "Two Worlds, One Sun." The gnomon was embellished with a spherical top or "nodus" that sort of represented the Sun in our mini-solar-system dial face. A daisy wheel-shaped secondary nodus was added partway down the gnomon to provide the sharp shadow edges needed for telling time with the dial when the Sun is low in the sky. We added two small mirrored segments on the face to allow us to see the color of the sky in each MarsDial image. The word "Mars" was etched onto the corners of the face in seventeen languages (more, if you count all the other languages besides English in which Mars is spelled "Mars"), acknowledging the astronomical contributions of Earth's ancient/indigenous peoples as well as representing the majority of the current written languages on our home planet.

On the MarsDial sides, in a tiny font too small to see in the images, even with our high-powered Pancam cameras, we added some stylized drawings of people, rockets, and rovers provided by chil-

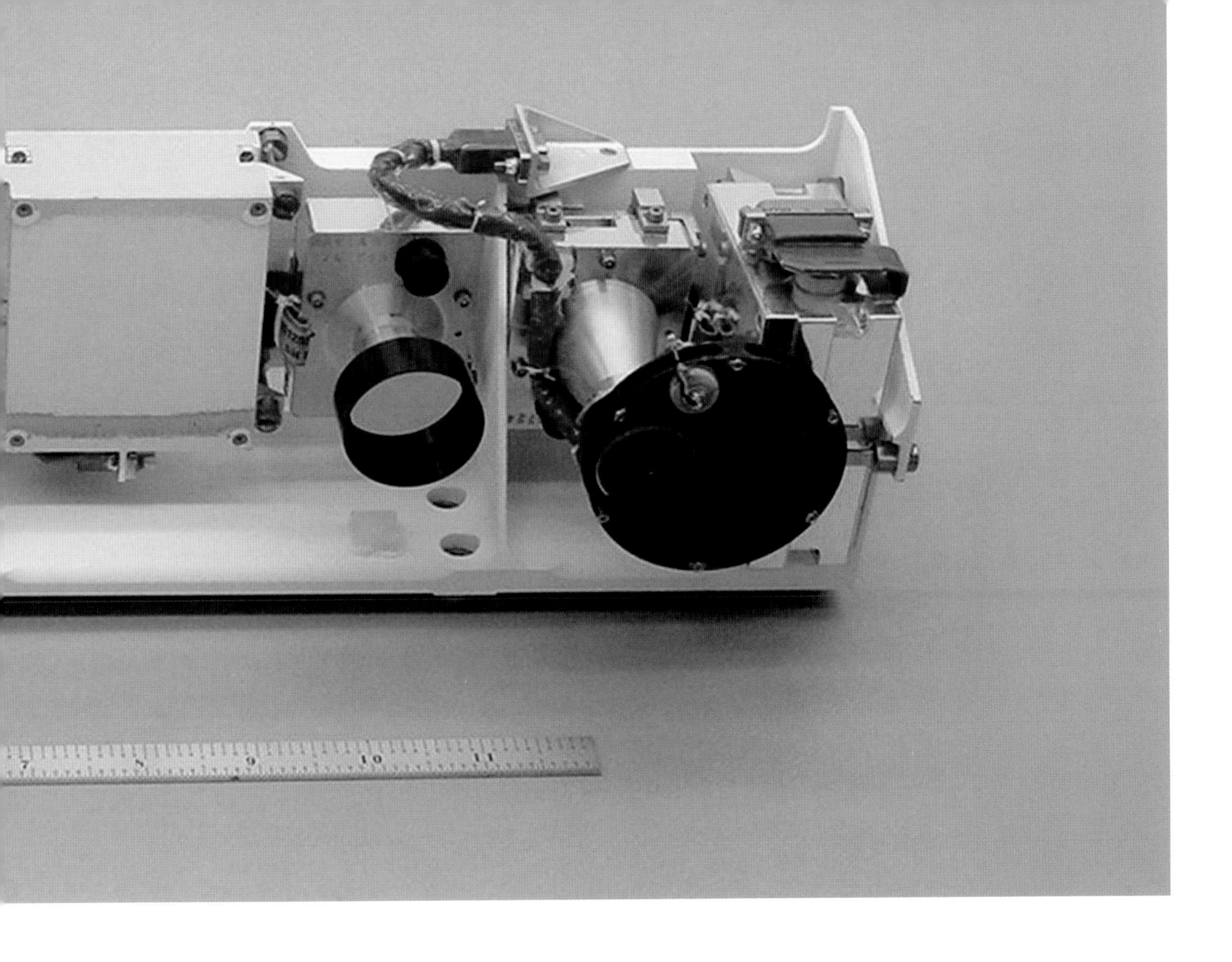

The *Opportunity* camera bar, before being mounted to the rover's mast. The Pancams and their filter wheels are the outer cameras; the Navcams are the inner cameras. The ruler is about 25 cm long. (L.WAYNE/JPL)

dren as well as an inspirational message for whatever future astronauts happen upon the rover and this curious little object mounted to its rear solar panel. The message says: "People launched this spacecraft from Earth in our year 2003. It arrived on Mars in 2004. We built its instruments to study the martian environment and to look for signs of life. We used this post and these patterns to adjust our cameras and as a sundial to reckon the passage of time. The drawings and words represent the people of Earth. We sent this craft in peace to learn about Mars' past and to prepare for our future. To those who visit here, we wish a safe journey and the joy of discovery."

The Expedition Conceived

On the afternoon of June 10, 2003, I watched a spectacular trail of flames and smoke rise into the clear blue Florida skies above Cape Canaveral launch complex 17B. *Spirit* was flying—hurtling—on the top of a powerful Delta rocket toward some destiny out in space. I was surrounded by dozens of colleagues who had played critical roles in getting the *Mars Exploration Rovers* to the launchpad, and hundreds more were watching from other viewing sites at the Cape and live on NASA TV. The looks on the faces I could see told me that many were experiencing something similar to me. My heart was racing, and I felt like dancing and throwing up at the same time.

This launch, and the launch of *Spirit*'s twin, *Opportunity*, about a month later, were the culmination of a crazy roller-coaster ride that we had been on for more than a decade. The idea of a robotic geologist roving across the rusty sands of Mars was the brainchild of a small group of planetary scientists and engineers lead by Steve Squyres in the late 1980s and early 1990s. Steve is a smart guy with outstanding scientific credentials, but that's only one dimension of his character. In some ways, he's like one of those eleven-dimensional cosmic superstrings that astrophysicists talk about—a unique entity vibrating to and fro at some insane rate and occupying some vast number of alternate realities all at once. Steve thrives on the thrill of exploring the new, and as we all learned while working with him, he's also one of those personality types that thrives on crises.

The package of scientific instruments that eventually morphed into what we called the *Athena* payload carried by *Spirit* and *Opportunity* was first proposed to NASA in 1995 as a set of experiments designed to be carried on a Mars lander to be launched in 1998. These instruments included a stereo color panoramic camera called Pancam (which had been under design and development for several years already, having been unsuccessfully proposed to NASA for their 1997 Mars lander mission), an infrared spectrometer called MINI-TES (for "Miniature Thermal Emission Spectrometer") based on one called TES that was on its way to Mars on the *Mars Global Surveyor* (MGS) orbiter, a soil sampling and water analysis instrument, and a meteorology package. There were also two German rock and soil analysis instruments and a microscope that would be carried by a tiny deployable rover called a Nanokhod that was also built by the Germans.

We wanted to answer some fundamental questions about Mars: Was there ever liquid water ponded (or "laked"? or "oceaned"?) on the surface? Is there still liquid water near the surface today? Was the atmospheric pressure higher in the past? Was the atmospheric composition different in the past? Was the environment, like that of early Earth's, one that would permit complex or even simple biologic or biochemical processes to occur, and is any evidence for these processes preserved on the surface today? Mars is on the solar system's short list of places to examine closely for evidence of life—past or present; the list also includes Jupiter's large moon Europa, which might have an ocean under its icy crust, and Saturn's large moon Titan, which has a thick atmosphere that is thought to be similar in some ways to the atmosphere of the early Earth. If Mars was Earthlike early in its history and if life is ever found to have developed there, it would be one of the most outstanding discoveries in the history of science. Such a discovery would provide support for those who believe that life is ubiquitous in the Universe. If conditions were favorable but life did not develop on Mars, however, then it will lend ammunition to those who believe that life on Earth is a rare or once-in-a-Universe quirk. In a sense, our mission would focus on Mars exploration as a way to test the necessity of the very existence of the word "exobiology," the study of life not on Earth.

Overall, our proposed 1995 mission, which we called MACS (for Mars Ancient Climate Surveyor), represented a strong set of scientific instruments—the strongest set of instruments that had ever been proposed to NASA for Mars exploration, we thought. However, as often happens in the space business, the NASA review panel and managers didn't agree, and instead opted for a competing mission judged to be superior to ours partly because it exploited a rare celestial mechanics opportunity to land near the planet's South Pole for less than the normal amount of energy.

We were disappointed, of course, but we knew we had a good thing going. We re-proposed the instruments in 1996 as part of a new mission for NASA's "better, faster, cheaper" Discovery small mission program. But we upped the ante: We had been excited about the possibility of doing the '98 lander mission, but what everyone agreed that we *really* needed to achieve our science goals was mobility, and that meant putting all of our instruments, not just the ones from the Nanokhod, on a rover instead of a lander. Getting a science payload like ours on a capable, mobile platform meant that we could sample the real diversity of rocks and soils and geology at a landing site rather than be constrained to whatever we could happen to reach from a stationary lander. It was an important part of the sense of "being there ourselves" that we really wanted to get out of this exploration effort. If any of us were taken out to some remote geologic field work site on Earth, for example, we'd have to move around, perhaps quite a bit, to get to and study the most interesting places and to figure out the geologic history of the area. Roving made scientific sense. And my camera team would be able to take much more interesting photographs.

We had to make some changes in the mission and instruments, many because of the "cheaper" part of "better, faster, cheaper." We planned for our rover to be launched to Mars in 2002 and carried on a lander similar to the design being used by NASA for what by then had become the 1998 *Mars Polar Lander* mission. We kept Pancam and Mini-TES and beefed up the capabilities of their mast. The Nanokhod was gone, but we kept its chemical and mineral spectrometers and microscope on the end of a strong new arm on the front of our bigger rover. We couldn't afford the soil chemical and water analyzer, so we replaced it with a very capable new instrument called a Raman spectrometer, which could provide detailed information on the soil and rock mineralogy. We added a miniature rock coring tool and a little sample canister onboard so that soil, rock, and core samples could be cached by the arm for a possible future rover to pick them up and return them to Earth. If our 1995 MACS proposal was good, this one, which Steve Squyres decided to name *Athena* after the Greek goddess of wisdom, was even better. We had cameras, spectrometers, and mobility. Our goals were focused on the most important questions that we could ask at the time about the history of Mars and the possibility of life on the Red Planet.

But again, we lost. This time it was a more bitter pill to swallow, though, partly because politics and fate worked against us. We had a strong science payload and a damned sexy mission—no one argued with that. While we were toiling away putting our proposal together in 1996, though, the entire landscape of Mars exploration changed around us. It was the fault of a rock—a meteorite, to be more precise—shaped like a big Idaho potato that had been found in Antarctica twelve years earlier.

Bugs from Mars?

Called ALH 84001 because it was the first meteorite cataloged from Antarctica's Alan Hills meteorite expedition in 1984, it's an unimposing little specimen that has had more influence on Mars exploration per gram of its mass than anything else ever touched by humans. The rock was special because it was one of a rare collection of meteorites known to have come from Mars based on analysis of small bits of trapped atmospheric gases found inside them. These rocks had been blasted off Mars millions of years earlier by asteroid or comet impacts and sent on trajectories that would eventually cause them to crash on Earth. One of these rocks, ALH 84001, became an instant international household celebrity in the summer of 1996 when researchers from NASA's Johnson Space Center in Houston, the same place where all of the *Apollo* Moon rocks are studied, announced that they had found evidence for ancient, preserved, fossil life *inside* this rock from Mars. Evidence of past life on Mars! It was stunning—not just to scientists but also to the general public who heard about it live on CNN from the president of the United States and in rapidly evolving forums and discussion groups that were cropping up on a new kind of communications medium called the Internet.

The evidence for life-forms having been preserved in ALH 84001 has turned out to be less compelling than most researchers first thought, and the verdict is still out on this issue within the scientific community. Were the anomalous shapes in the rock really ancient Mars bugs? Either way, the most important message from ALH 84001 was that it provided strong evidence that conditions at or near

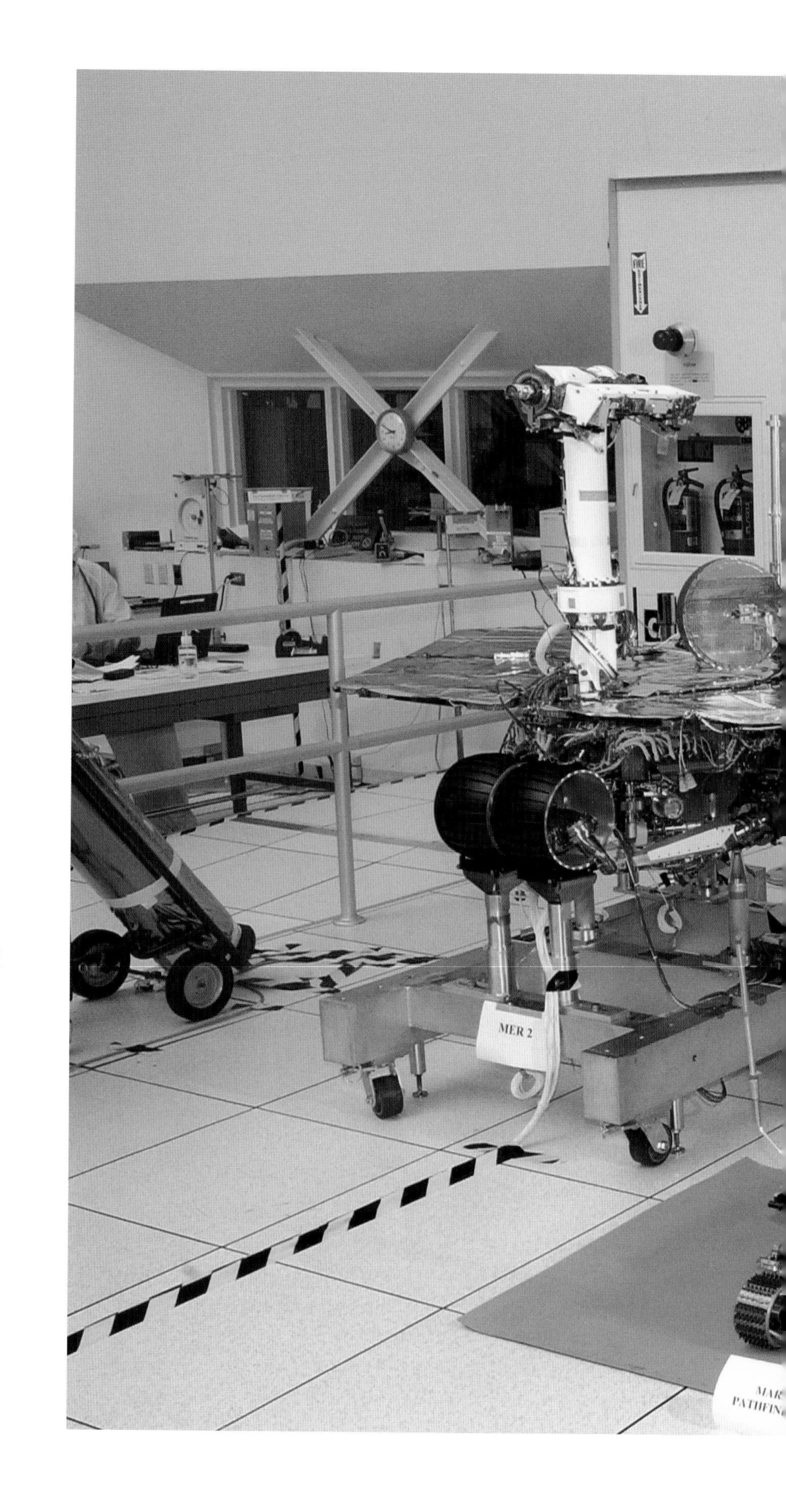

Three rovers: *Spirit* (left), *Opportunity* (right), and the *Mars Pathfinder* flight spare *Marie Curie* (foreground). (NASA/JPL)

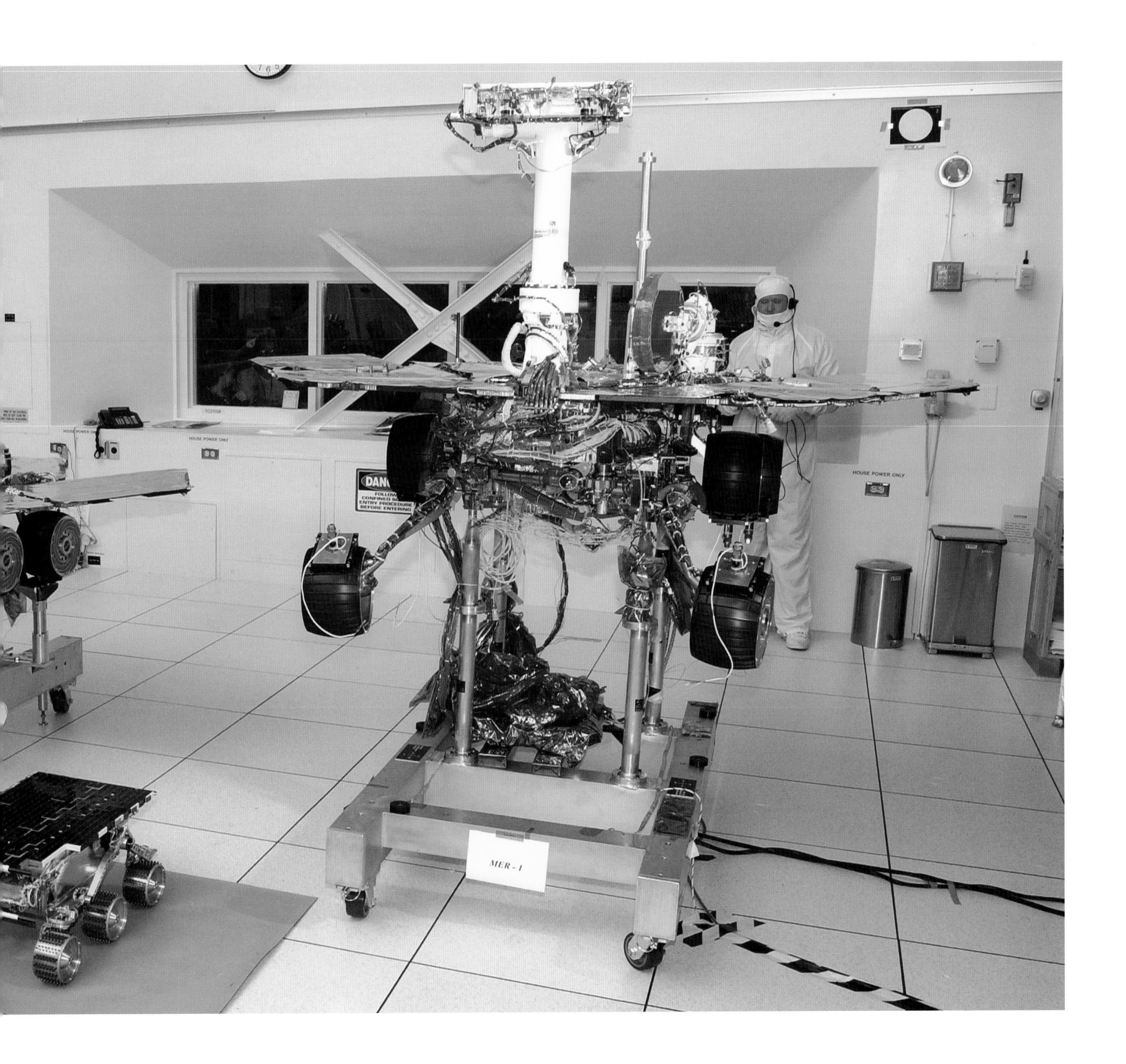
HOUSE POWER ONLY
ENTRY PROCEDURE
BEFORE ENTERING
HOUSE POWER ONLY
MER - I

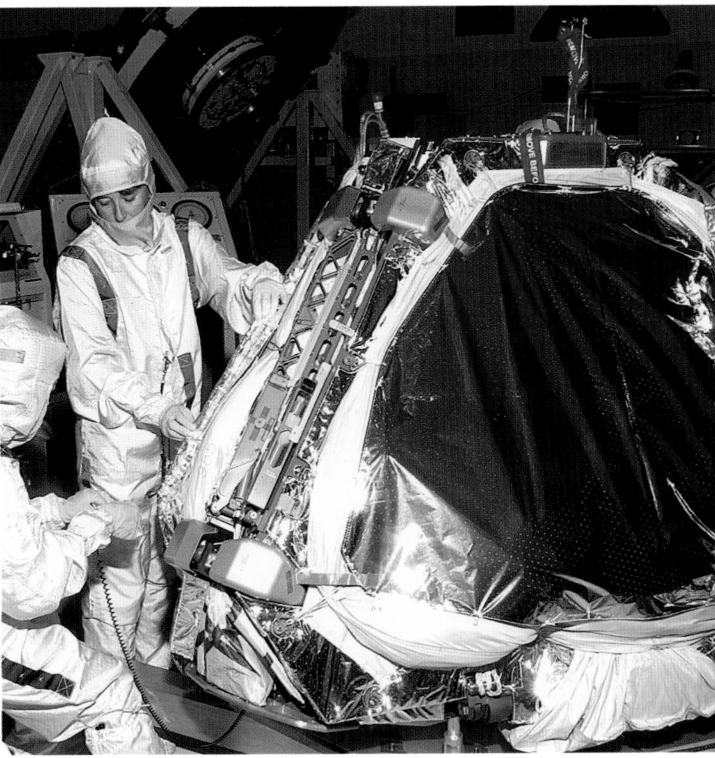

the surface of Mars long ago were much more Earthlike than they are today. The rock is from Mars—chemical data indicate it, and there is little argument among scientists about it. Certain kinds of minerals were found inside the rock that demonstrate that *liquid water* flowed through it. The rock was *heated* by some process—perhaps volcanism, perhaps impact, perhaps both. And the rock contains traces of *organic molecules*, some of which do not appear to be the result of contamination after it arrived on Earth. Liquid water, heat sources, and organic molecules are "the Big Three" on the list of requirements for "Life As We Know It" on a planetary surface. Overnight, the Mars exploration program became a media sensation again, and the prospects for life on Mars gained significant scientific credibility.

Ironically, this was the death knell for our *Athena* Discovery rover proposal. Public, congressional, and presidential administration interest in Mars fueled a big increase in NASA's Mars exploration budget and created new opportunities for Mars missions. That meant that our proposal to Discovery—a program designed to generate science missions to planets, moons, asteroids, and comets all over the solar system—was suddenly oddly out of place. Expecting the Discovery program to pick up the tab for a Mars mission when the Mars program itself was expecting a pile of cash soon would be naive. "Why don't you try to do this mission from within the Mars program's budget?" would be a likely reply. And so, in effect, that's what NASA and the Discovery program review panel asked us to do with our *Athena* rover.

Try and Try and Try Again

Battle weary, the team dusted off our *Athena* proposal again in 1997 and submitted it to NASA for a new mission to Mars, launching in 2001. By this time, NASA had signed on to the idea of roving

Mars with cameras (spectrometers, and all the rest). Everyone was being asked to propose instruments not for a lander, but for a new Mars *rover*. Not to our surprise, the rover that would be built by JPL had remarkably similar capabilities and requirements to the one that we had proposed to the Discovery program the previous year.

This time, we thought, we'd get it. We *knew* that our payload would work with the rover design that NASA was describing because our team (which had included many colleagues from JPL) was instrumental in designing the rover in the first place. And we had a public salivating with Mars mania, having swallowed ALH 84001. But our job this time required some significant sacrifices. NASA was given more money by Congress for this new mission, but most of it would have to go toward building the rover, launching the rocket, and operating the spacecraft on Mars. There was precious little left for the science instruments—a smaller fraction, in fact, than what we had assumed for our Discovery proposal. This meant that we had to make some painful cuts in our original *Athena* payload. The worst two cuts involved Pancam and the Raman spectrometer. We had been carrying a Pancam design that dated back to a concept first worked on by Steve Squyres and colleagues at Ball Aerospace in 1993. It was a true digital panoramic camera, using essentially a long slit that was rotated around the inside of a canister with optics that were optimized to build up huge, wide images. The design was sleek and a prototype camera that was built took beautiful images, but it was expensive to build, partly because it was very different from the other cameras that the rover would carry for navigation and hazard avoidance (which were more similar to traditional digital cameras that you can buy off the shelf). The more cost-effective but painful thing to do was to abandon the original Pancam design and switch to one that was more like the rest of the rover's cameras. Raman spectrometers are complex instruments and as a result are rather expensive. It would have been worth the expense in terms of the science data that it would have returned for us, but we just couldn't fit it into the "cost box" that NASA required. We didn't toss it outright, but we had to carry it as an optional payload component in our proposal that we hoped NASA could find a way to pay for later. No one was particularly optimistic about that possibility (which didn't come to pass, ultimately), but it was the best that we could do.

Our proposal was at last accepted by NASA. Unfortunately, shortly after we were selected in 1997, NASA and JPL announced that a combination of engineering problems and funding shortages in the Mars program would mean delaying the launch of what was now being called *Mars Geological Rover* until the next Earth-Mars launch opportunity in 2003, with a landing in 2004. This was a disappointment, of course, but if it meant a more robust rover and lower-risk landing system, then it was clearly worth it.

Then, in the span of three months in late 1999, NASA lost the *Mars Climate Orbiter* and *Mars Polar Lander* spacecraft. The *Orbiter* burned up in the martian atmosphere due to poor team communications and a resulting human navigation error based on—of all things—a silly conversion mistake between English and metric units. *Mars Polar Lander* crashed into the surface, probably as a result of a software error during the complex sequence of landing events. The Mars program was thrown into turmoil by these highly visible failures, and public confidence in NASA was shaken, yet again. The entire future of Mars exploration, in fact, was placed under the gun for close scrutiny by several blue-ribbon (and congressional) investigative committees. As a result, our 2003 *Athena* mission was also scrubbed, pending a restructuring of the NASA and JPL Mars exploration programs. Our roller-coaster ride to the Red Planet was getting more and more dizzying. In the spring of 2000, our hopes for roving on Mars were grim.

In the late spring and summer of 2000, though, our hopes were raised again. NASA's restructuring activities had identified several options for recovering the program in the 2003 launch opportunity. Among them were an orbiter with a comprehensive remote sensing science package and a rover car-

LEFT
Technicians fold up *Spirit* inside her protective flower-petal-like cocoon lander, just before launch. (NASA/JPL/KSC)

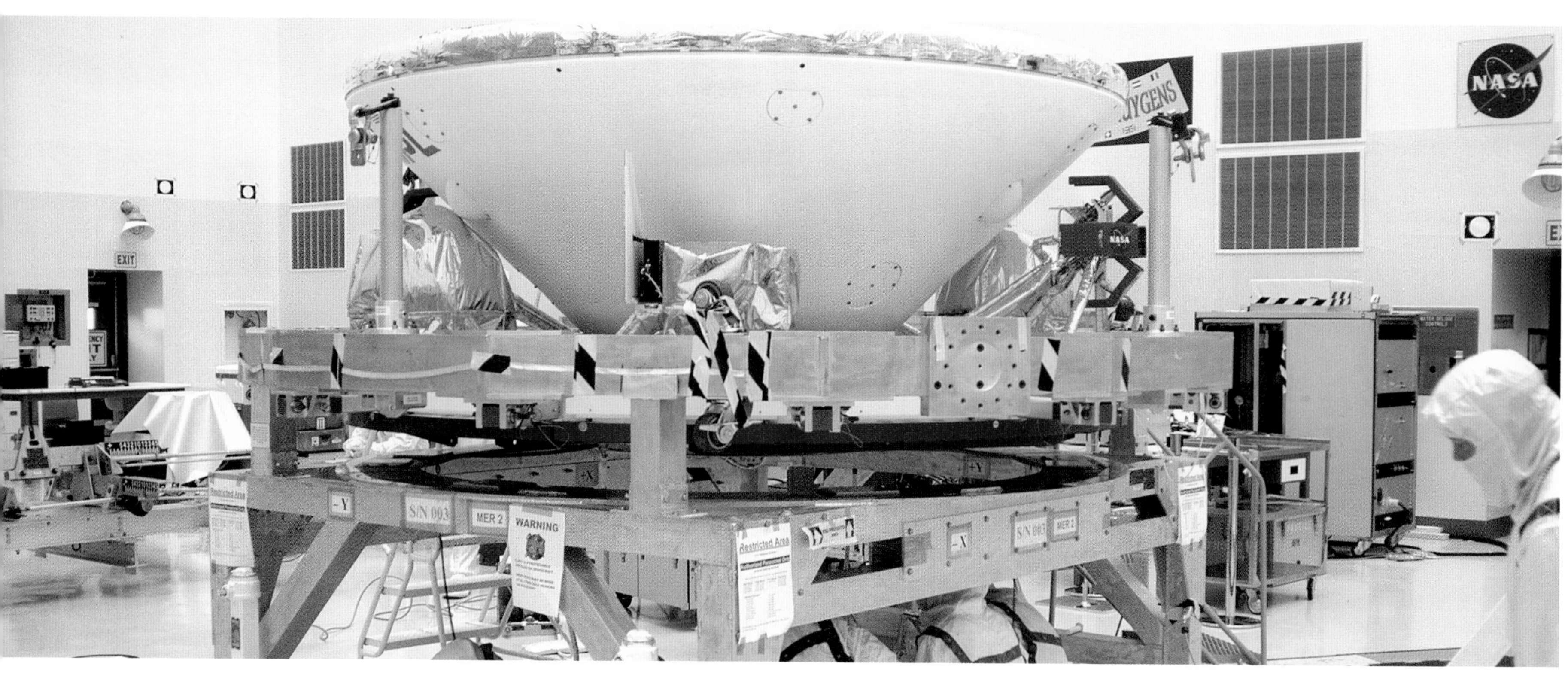

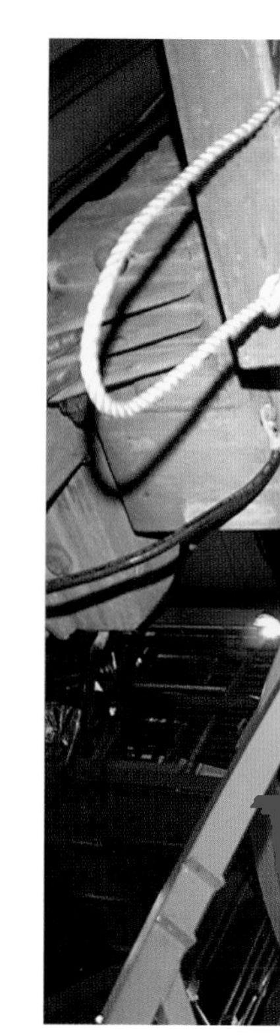

rying our *Athena* science payload, but this time landing on Mars using a parachute-and-airbag system inspired by the successful 1997 *Mars Pathfinder* landing system. Engineering and science studies were conducted, documents were written, and presentations were given on both coasts as team members raced between Washington and Pasadena making the case to save the rover. Hope turned to anticipation as NASA announced in July 2000 that they had selected our rover mission to lead NASA's first Mars surface exploration mission of the twenty-first century. But there was still more: In order to reduce mission risk and to take advantage of the rare close alignment of Earth and Mars in 2003, NASA Administrator Dan Goldin had decided that it would be more prudent to send *two* identical rovers to Mars in 2003, each of which would carry our *Athena* instruments to a different landing site. This would represent a return to the old ways that NASA used to reduce risk in interplanetary missions (twin *Voyager*s, *Mariner*s, *Viking*s), but it was still a surprise to everyone involved. I don't know for sure how long Steve Squyres and the JPL management took to make the decision when they were asked if the team could build two rovers and two sets of instruments. I'd bet milliseconds. I'm not sure many of us realized at the time what would really be involved. It was a euphoric time, and our dreams of roving on Mars were probably clouding everyone's judgment. In the end, it is ironic that the tortuous path we had to take to this exciting and unexpected outcome is remarkably close to what we originally proposed to do back in 1995, except using two rovers and a now battle-worn science team. That Squyres was the father of the twin rovers *Spirit* and *Opportunity* there is no doubt. Their mother, though, like so many other mothers of invention, was a mélange of necessity, inspiration, perspiration, and even some plain old dumb luck.

One of the keys to success was the extremely close cooperation between scientists and engineers at every level and throughout every phase of our project. It is often said that engineers and scientists just don't play well together. In some space projects that I've worked on, this has sure

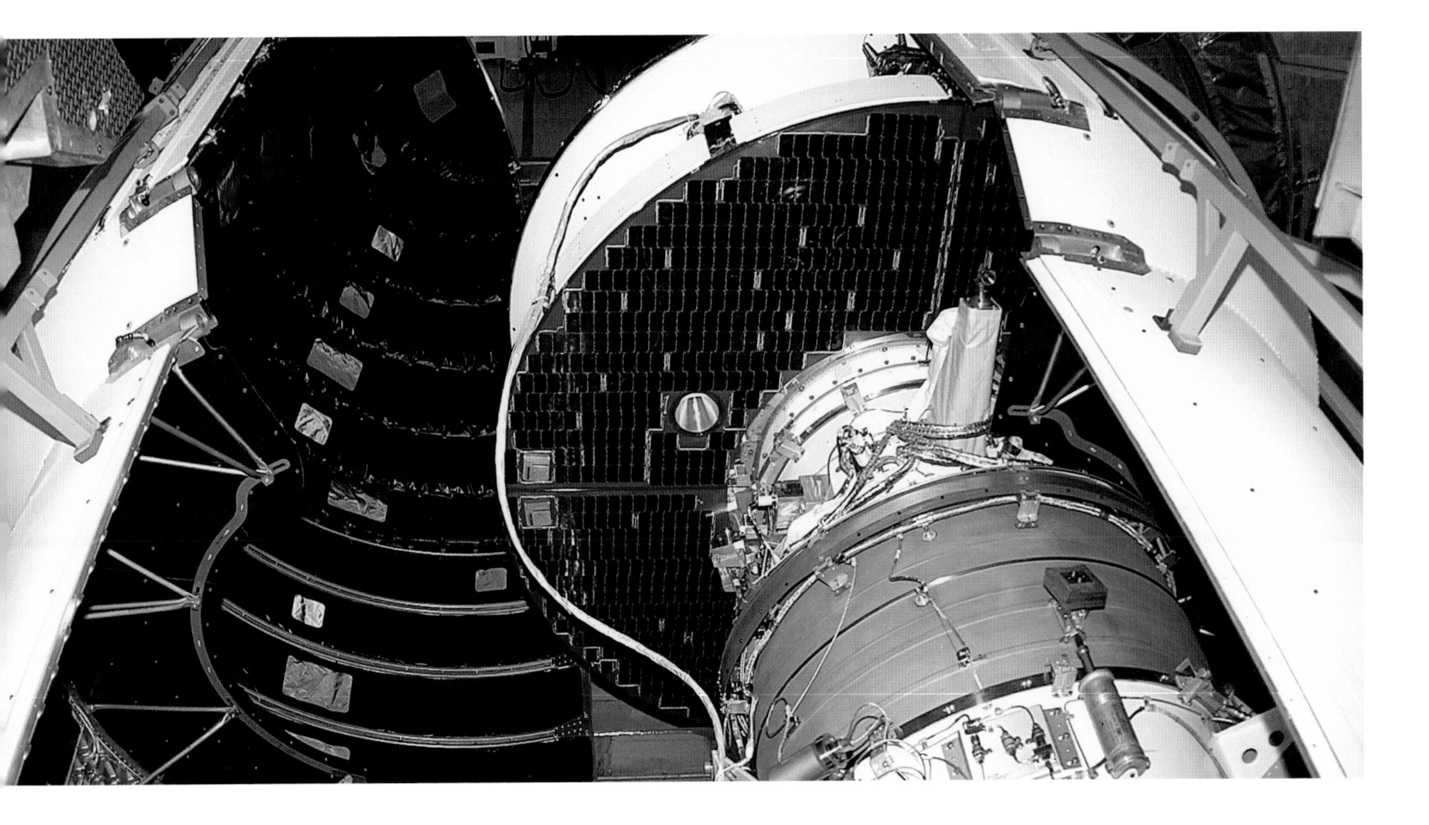

seemed to be the case. The job of the scientists, we would be told by some Engineering Team leader type, was to Define The Requirements that the instruments needed to meet. For example, what kind of resolution did we need on a camera, or what range of colors did we need to detect on a spectrometer? Then we were to essentially throw those requirements over a cubicle wall to a waiting tribe of engineers who would then go off and build instruments to deliver them. Sometime later, if everything went well, our data would come flying back over the wall and then we could go off and write papers (because that's what scientists do). Engineer friends have told me about the experience from their perspective. The job of the engineers, they would be told by some Science Team leader type, was to build perfect instruments that could make noiseless measurements using minuscule amounts of power, and to endow the instruments with the flexibility to be reprogrammed and reconfigured, if necessary, at any point in the process—even after launch—as the scientific goals were modified on the fly.

A dual science/engineering mentality on big space mission projects can lead to misunderstandings, disputes, and even bigger problems if left unchecked. Happily, scientists and engineers on our rover project formed a sense of team unity right from the start. The tone was set by the level of cooperation that formed between the leaders of the different science, engineering, and management teams. There were science team meetings where engineers and managers were invited and encouraged to attend, and engineering and management reviews where scientists were similarly welcomed. Scientists were encouraged to attend "Flight School" to learn about the rover's systems and capabilities in great detail, and engineers were encouraged to attend special science lectures to learn about what scientists needed to measure and why. Apparently, if you try hard enough and have the right leaders, scientists and engineers can play well together.

LEFT TO RIGHT
Opportunity, safe inside her lander and cruise stage, being prepared for mating with the third stage of the Delta rocket. (NASA/JPL/KSC)

Opportunity atop the Delta second stage, just before closing the rocket's fairing. (NASA/JPL/KSC)

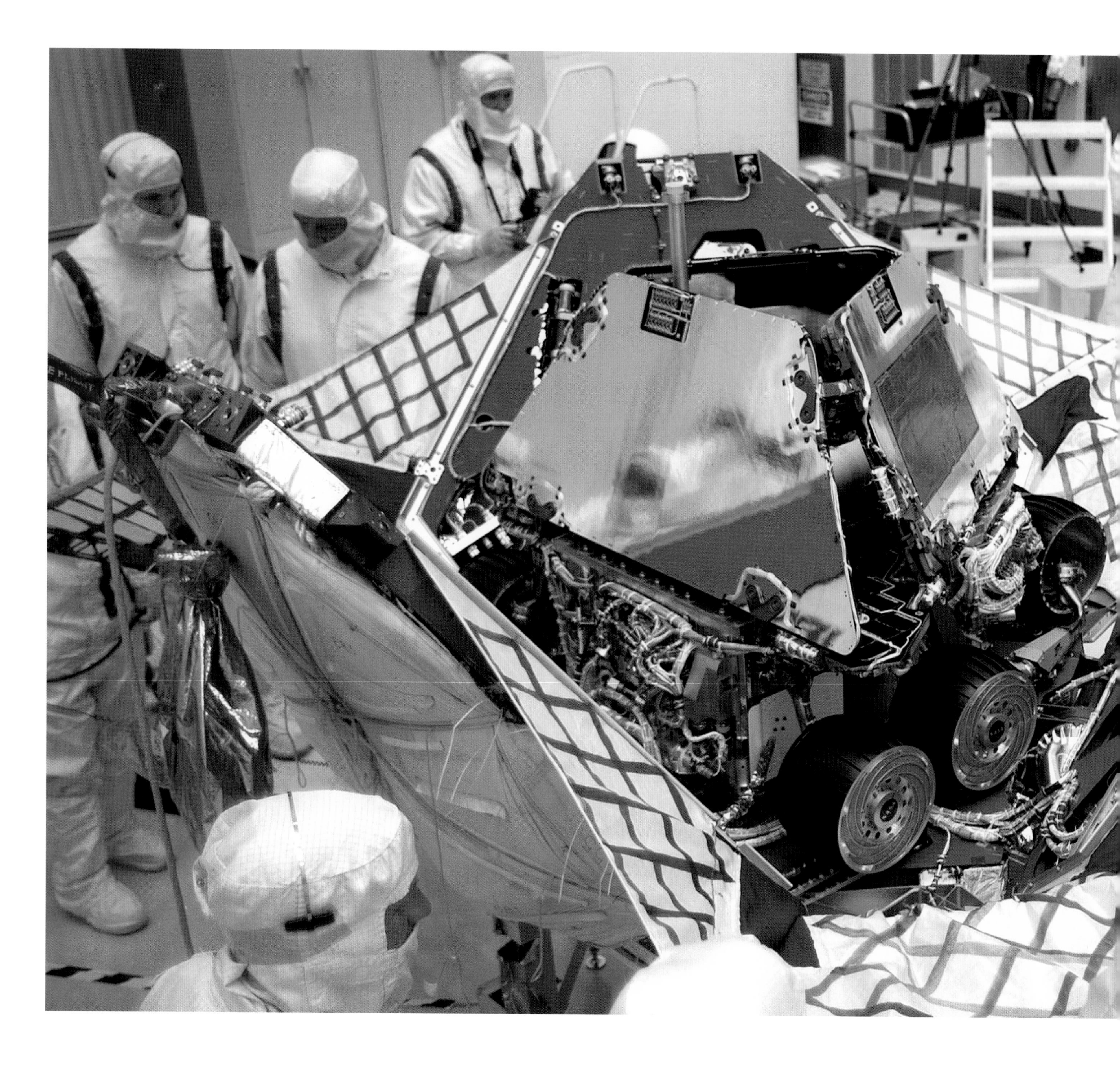

Technicians fold up *Spirit* inside her protective cocoon lander, just before launch. (NASA/JPL/KSC)

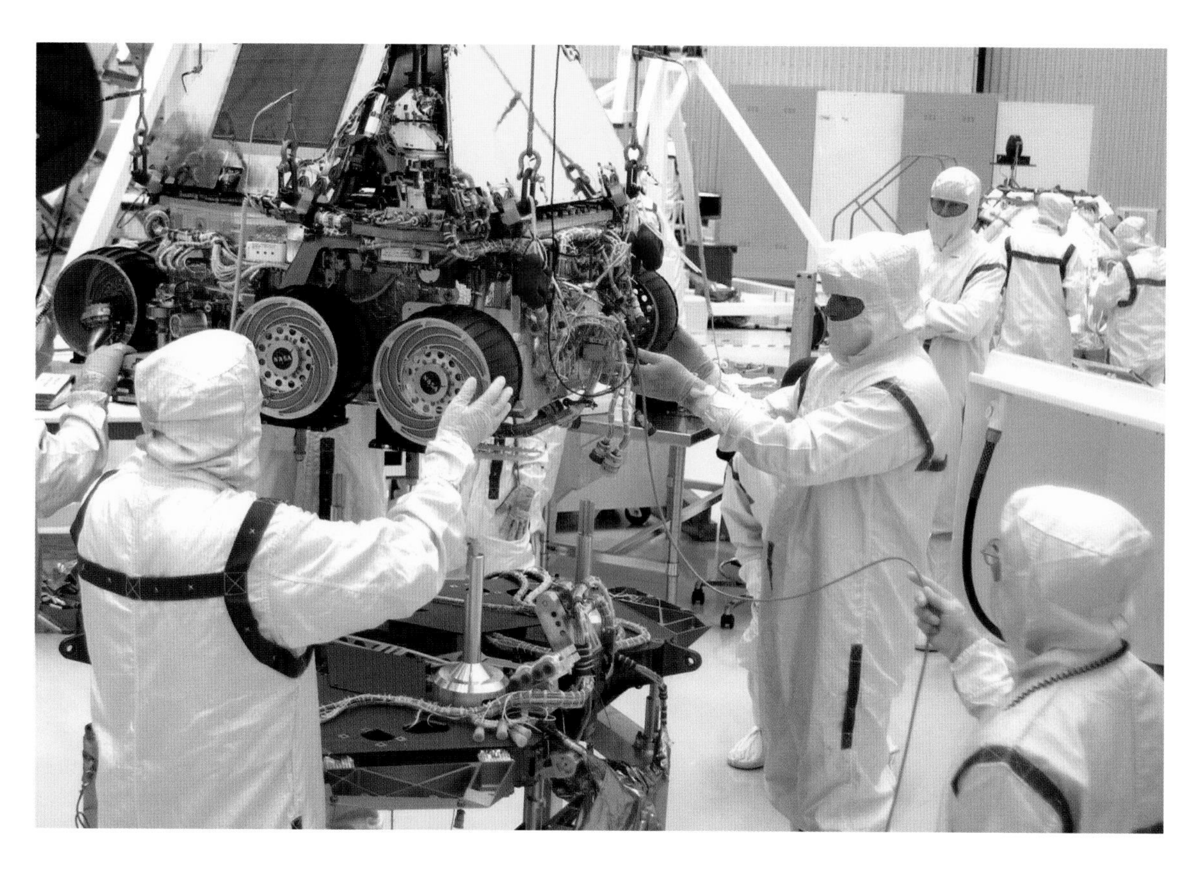

LEFT TO RIGHT
Opportunity being lowered onto her lander's base petal prior to launch. (NASA/JPL/KSC)

Spirit rolling over ramps to test her mobility and maneuverability in the final assembly and test laboratory at Kennedy Space Center. (NASA/JPL/KSC)

Opportunity, folded up safely inside her lander, getting attached to the backshell during final prelaunch preparations. (NASA/JPL/KSC)

Off and Running

So we played. The three-and-a-half-year period from when NASA gave the go-ahead to build the twin rovers to the summer of 2003 was a crazy sprint toward a deadline set by Newton's laws of celestial mechanics. About every twenty-six months, the Earth and Mars are lined up just right, so that a rocket can use the minimum amount of energy to travel from here to there. Using the minimum amount of energy means using the minimum amount of rocket fuel, which means using the smallest possible rocket, which means the minimum amount of money. NASA is strongly motivated to use those biannual opportunities for Mars launches, and the next one coming up for the rovers was in a "window" extending from June to July of 2003. If we didn't get the rovers to the launchpad in time for that window, we'd have to wait another twenty-six months for the next opportunity. But it wasn't that simple: Because of a rare close alignment of the Earth and Mars in 2003 (the planets were closer that summer than they had been in 60,000 years, and closer than they would be again until the year 2287!), 2003 offered the *best* launch window for a long time. If we missed that window and had to wait until 2005, we'd need more energy, more rocket fuel, bigger rockets, and more money—or we'd have to completely redesign the rovers to make them much smaller. Clearly, if we didn't launch the rovers in the summer of 2003, we'd probably never launch them. The clock was ticking.

By early 2002 we had settled on our near-final payload instrument design. These rovers were going to be the best-equipped robots ever sent to Mars.

Each rover would carry cameras—and lots of them. Each rover was outfitted with nine separate cameras, and a tenth was carried by the lander to take pictures of the landing site during descent. Of the nine rover cameras, two pairs are very wide-angle (fish eye), black-and-white ste-

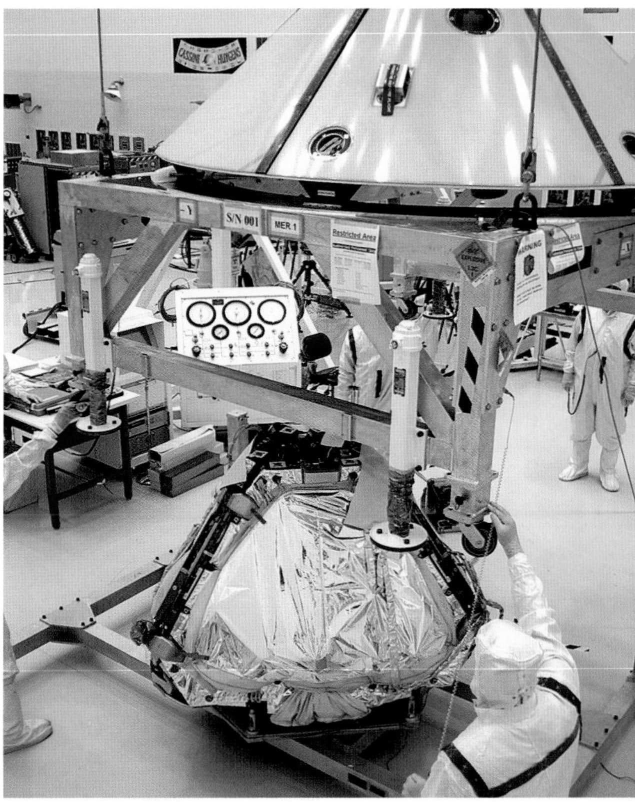

reo cameras called Hazcams (hazard avoidance cameras) because they are mounted low to the ground at the front and rear and were designed to help characterize potential hazards while the rover is driving. Two more pairs of cameras are mounted atop a 1.5-meter-tall mast at the front of each rover. One of these pairs is a set of wide-angle, black-and-white stereo cameras called Navcams (navigation cameras) because their main purpose is to provide quick stereo coverage of lots of terrain around the rovers for navigation. An energetic young scientist-engineer hybrid named Justin Maki from JPL was responsible for overseeing the testing and operation of the descent cameras, Hazcams, and Navcams, collectively called the "engineering cameras." The two remaining cameras on the mast are the Pancams, which were designed to obtain very high resolution images of Mars in many different colors, ranging from ultraviolet colors somewhat "more blue" than the human eye can perceive to infrared colors somewhat "more red" than human color perception. Back when he transitioned from a camera honcho to our overall project science leader, Steve Squyres gave me the responsibility of overseeing everything having to do with the Pancams. It felt like getting the keys to someone's favorite car (and a $13 million car at that!). Of course, I worried about whether I was up to the task. Steve was taking the keys to a bigger and even more expensive car, though, and it was heartening to learn that he was also worried about whether he was up to *that* new task. The ninth rover camera is a small microscopic imager on the rover's arm, designed by Ken Herkenhoff and colleagues from the U.S. Geological Survey (USGS) in Flagstaff to take black-and-white images of regions only about three centimeters across at a scale comparable to a geologist's hand lens (because, after all, you'd never find a geologist out in the field without a hand lens). Ken is an expert geologist and proved

to be an excellent source of patience and expertise for tackling problems that came up with all of the cameras. Images from the nine rover cameras and the descent cameras would provide the information we'd need to drive and to explore the geology and topography of each landing site. Pancam would be providing the only color pictures from the rovers and would be the first moving set of (essentially) human eyes on the planet.

Each rover would carry spectrometers—three of them. One is the Mini-TES, designed by Phil Christensen and colleagues at Arizona State University and mounted deep inside the rover. Phil is the charismatic leader of the TES team, which runs an instrument that is on the MGS orbiter and is the big brother of Mini-TES. Phil assembled an amazing concentration of the world's best infrared engineering and science specialists at his lab in Tempe. Mini-TES uses the Pancam mast as a periscope to detect infrared (heat) radiation from rocks and soils. Scientists then use that information to determine the kinds of minerals that are there. The two other spectrometers are both mounted on the rover's arm. One is called the APXS, for Alpha Particle X-ray Spectrometer, and it measures the abundances of different elements in the soil (silicon, iron, calcium, aluminum, etc.). The APXS is a German instrument, the brainchild of Rudi Rieder and Ralf Gellert, a yin and yang duo of geochemistry and instrumentation expertise from the Max Planck Institute in Mainz, Germany. The APXS is an enhanced version of a similar instrument that this same team flew on the *Mars Pathfinder*'s little *Sojourner* rover back in 1997. The final spectrometer is called a Mössbauer (MB) spectrometer (pronounced, as best as I can figure, with the first part like "murse" and the last part like "bower"—it's the last name of the physicist who invented it) and is designed specifically to identify iron-bearing minerals that are known to exist on Mars and are responsible for the planet's reddish color. This was also an instrument contributed to NASA by the taxpayers of Germany. Getting one of these instruments to Mars had been a lifelong dream of its creator, Göstar Klingelhöfer, a clever and cheerful colleague who is an expert in iron minerals.

Each rover would carry a grinding/drilling device, called the rock abrasion tool (RAT for short), also mounted on the rover's arm and designed to brush or scrape dusty coatings or rinds off of rocks so that the spectrometers and microscope could investigate the insides of the rocks themselves rather than just the dirty, grungy outsides. The RAT was designed and built by a crew of young and clever engineers at a small Manhattan company called Honeybee Robotics, which is run by a softspoken but passionate robotics engineer named Steve Gorevan. You'd swear by the demeanor, accent, and haircut that Steve was a long lost Kennedy brother. Getting inside the rocks was a major science goal for our mission, and the Honeybee's RAT turned out to be crucial for making that happen.

And finally, each rover would carry several other instruments and systems used for science measurements. A set of magnets built by a group of Danish colleagues was used to trap or deflect different components of the magnetic airborne dust. The rover wheels were used for trenching and rock-scraping experiments. And a variety of calibration targets provided essential reference information for the cameras and spectrometers, including of course the first thing to photograph: our charming MarsDial.

LEFT
Solid rocket boosters being mated to the Delta II rocket for *Opportunity*'s launch. (NASA/JPL/KSC)

FOLLOWING SPREAD
The second stage of a Delta II rocket being moved to the top of the launch tower on Pad 17-B, Cape Canaveral Air Force Station. (NASA/JPL/KSC)

The Boeing Delta II rocket, carrying the *Spirit* rover on Launch Complex 17-A, Cape Canaveral Air Force Station, Florida. (NASA/JPL/KSC)

Killing a Camera

Perhaps the most grueling and intensive activity that the camera team went through prior to launch was a series of laboratory measurements during most of 2002 and the first part of 2003 that were carefully designed to test and calibrate the instruments. First by themselves, then after assembly on the rover, each camera was subjected to the shocks of launch and landing and the same kinds

NASA
USAF
BOEING
DELTA
MER-A
45TH SPACE WING
40 ASTRA

NASA
USAF
BOEING
Sverdrup
DELTA
MER-A

Spirit begins her journey to Mars. The rover was launched on June 10, 2003 at 1:58 p.m. (NASA/JPL/KSC)

of extreme cold and near-vacuum conditions that they would encounter on Mars. In fact, to play it safe, we had to prove that the instruments would work in even harsher conditions than the ones we expected to encounter on Mars. The only thing we couldn't simulate—though we would have had we been allowed—was coating the cameras with fine-grained, reddish dust like they'd see on Mars during the planet's famous dust storms. A few dozen of us—scientists and engineers from the team, as well as a handful of wide-eyed young Cornell students—spent months working in shifts to plan and carry out twenty-four-hour-per-day-seven-day-per-week tests inside cold, loud JPL basement labs surrounded by vacuum pumps and test chambers. We carefully measured how the cameras performed and pointed them at rocks and other targets of known characteristics to make sure we had the accuracy and sensitivity that we'd need on the surface of Mars.

It was certainly not the glamorous side of space exploration, but to take the photographs and do the science that we eventually wanted to do with these rovers, this kind of grunt work ahead of time was essential. We took pictures of a Starbucks Frappuccino bottle (empty) when we needed a green imaging target, and mug shots of one another with the flight cameras. We practically lived on corn chips and the wonderful smoky salsa at Doña Maria's, a little Mexican restaurant not far from JPL. Many of us had to take special JPL training classes to be allowed to handle spaceflight hardware or to use the calibration facilities and their safety rules. There was a separate team full of (trained-to-be) nervous Quality Assurance engineers who watched every move related to the flight hardware.

Some of the rules were strange. My favorite was the one that insisted that if two people were working alone in a lab and one was overcome by fumes, the other should run and get help *before* removing the victim from the room. Luckily, we never had to test breaking that rule. The calibration tests culminated in a series of thermal vacuum measurements where each complete, fully assembled rover was placed into a huge vacuum chamber and pumped down to Mars temperatures and pressures. We were living and working with machines that would soon be roving on the surface of another world.

I'm kind of a control freak normally, but I tried to be even more of a freak during all of this testing of the cameras. We had to build and test more than forty different cameras in only a few months (twenty cameras on the two rovers that would go to Mars, more for the two rovers that would be used for testing on Earth, and all kinds of spares for both). I was pretty grumpy most of the time, and didn't feel like I was making a lot of friends. Once, during the summer of 2002, it looked like things had gone totally haywire. A camera died.

We were working through the testing of *Spirit*'s Pancams, trying new setups to get the right data and writing new software on the fly to acquire and process the images as efficiently as possible. All of a sudden, after swapping some cables around and switching from one particular control computer to another, one of the cameras (serial number 103, *Spirit*'s right eye) started giving us nothing but zeros. Millions of them. Black as black. We tried restarting it, reconnecting it, swapping back the old computer. Nothing. It was dead. We thought for sure that one of our cable swaps must have fried something in the electronics. This was a nightmare. We were on such a tight schedule—no cameras, no launch. We could work our butts off to get one of the spare cameras ready instead, and maybe we would make it, but there were still only twenty-four hours in a day and only so many days until the launch window closed.

Luckily, thankfully, happily, miraculously, after calming down for a few hours and thinking about it a little longer, three of the team's hardest workers and most inspired problem solvers figured out the problem. Cornell team members Heather Arneson and Miles Johnson (who had started working with us as undergraduates), along with lead JPL camera systems engineer Mark

Spirit hurtles into space; June 10, 2003.
(NASA/JPL/KSC)

Schwochert, realized that the camera had been so cold that with the particular settings that we had used, sending zeros was exactly what it should have been doing. We tweaked the settings and, *voilà!* Number 103 came back to life, as if it had never left us. We all breathed a huge sigh of relief. I apologized to everyone I remembered swearing at over the course of that miserable night. It's odd, but the whole calibration process was insanely tense. I hope I never have to do anything like that again in my life.

Lofty Goals

What was the point of nearly working ourselves to death and spending more than $850 million of hard-earned taxpayer money on such a project? Because it's inspirational and educational and probably one of the coolest projects my colleagues and I will ever work on in our careers? Sure, but it's more than that. We're trying to do at Mars what all scientists try to do in general: make observations and measurements to test specific hypotheses, chew on the data for a while, revise or abandon some hypotheses and form new ones, then go out and make some new observations to test those. This incremental, communal process of testing and refining hypotheses is what scientists do. It's what leads to knowledge and, ultimately, understanding.

We wanted to use the rovers and their instruments to answer questions about Mars that were formed, revised, and argued about over the past few decades based on results from telescopes, meteorite studies, and missions like *Mariner, Viking, Mars Global Surveyor,* and *Mars Odyssey.* For example, some scientists think that Mars once had a thicker atmosphere early in its history, making temperatures warmer for long periods of geologic time. Others have countered that any thick atmosphere must have been lost very early in the planet's history—blown into space by huge cosmic impacts or slowly "eroded" away by the incessant bombardment of the solar wind. Some claim that Mars once had abundant liquid water flowing across its surface, possibly for long periods of time and in standing bodies like lakes or oceans. Others claim that the only compelling evidence for water indicates that it was short-lived on the surface and then quickly seeped back underground or "escaped" to space. There were two sides to every argument coin: (a) The surface of Mars has been extensively modified and altered by water and heat from volcanoes; or (b) Wind and impact cratering, not water and volcanoes, can explain what we see on Mars. (a) Mars has always been a sterile world; or (b) There is evidence for past or even present life on Mars. (a) Mars was like the Earth; or (b) Mars has been mostly dead, geologically, like the Moon.

While we got the rovers and their instruments ready, the debates about Mars went on, fueled by new data and new discoveries, as well as by the public's apparently insatiable appetite for learning more about the place in the hope that, by the process of exploration, we will learn new things about ourselves and our home planet.

Site Selection

We wanted to test many of these hypotheses with the instruments on *Spirit* and *Opportunity* by seeing firsthand the geology and mineralogy of two very special landing sites. Choosing just two places to land is a hard thing to do on such an interesting planet like Mars (imagine trying to choose just two places to explore to "discover" everything you could about Earth). The rocket and mission controllers could aim the landing spot to be within a long, skinny oval (called the "landing ellipse") about 100 km wide east-west and about 20 km wide north-south. This is pretty astounding accuracy: hitting a

100-km-wide bull's-eye from a total distance traveled of 500,000,000 km. I've heard this compared to hitting the torch on New York City's Statue of Liberty with a baseball thrown from London.

However, there were both engineering and science limitations preventing us from putting those landing ellipses anywhere we wanted on Mars. For example, the rovers are solar powered, so we couldn't land them in places too far from the equator, where the Sun is lower in the sky and the power levels would be too low. The landing system relied on a parachute to help slow us down, so we couldn't land above a certain elevation because there wouldn't be enough martian air for the parachute to slow the lander before it reached the surface. The parachute-and-airbag system had to be able to work within predictable performance limits, so we couldn't land in a place where the winds would be too severe or the daily weather couldn't be predicted reasonably well. We didn't want to land in a place where it was too rocky, because of the potential for the airbags to be punctured during landing, or because of the potential obstacles that could prevent us from driving around effectively. And we didn't want to land in a place that was just flat and boring, because then there might not be enough interesting science related to our mission goals.

It is perhaps not surprising that we found ourselves having to make a trade-off between scientific interest and landing safety. The safest places to land were often the flattest, most "boring" sites with the least compelling scientific potential (like the *Viking* lander sites, which were selected primarily based on their landing safety assessments). In contrast, the most scientifically exciting and interesting places to land on Mars are often some of the most dangerous or least accessible: canyons, mountains, channels, polar caps.

NASA and JPL went out of their way to make the site-selection process inclusive, holding a series of public workshops over the course of three years to decide where to send *Spirit* and *Opportunity*. The first workshop had around 200 scientists contributing to it. Not surprisingly, there were about 200 different landing sites proposed. Many of these were winnowed out quickly because of engineering or safety considerations. Over the course of subsequent workshops more difficult scientific debates were held to try to rank the few dozen serious contenders. We held our own debates within the rover science team, trying to rank our own favorites to present to the larger community at the workshops. Sometimes the process was frustrating, especially when the engineering or other constraints would change in response to rover or mission design changes. Often some good science contenders were tossed out of the field, and sometimes others were revived.

My favorite was a site in a place called Melas Chasma, part of the large (5,000-km-long) canyon system known as Valles Marineris. I thought of it as landing in the Grand Canyon on Mars. Melas is in a part of the canyon system with layered deposits in the walls and evidence that there are different kinds of water-related sedimentary rocks exposed in those layers. It would be a visually stunning sight, looking up at steep, layered walls while driving through the canyon. Alas, the site was tossed out because some models of the winds coming down those canyon walls predicted that they could be hazardous to the landing system. It was frustrating to lose the site because of a *model* of what might happen, but there was no other way to assess the risks because we don't have any actual wind speed measurements from Melas. If only we could send a lander or rover to the surface there to monitor the weather, we'd know for sure. A NASA catch-22.

Eventually, the site-selection process whittled the list down to the top four potential places to land our rovers on Mars: Meridiani Planum, Gusev Crater, Isidis Basin, and Elysium Planitia. Meridiani Planum ("the plains of the meridian," near 0° longitude on Mars) is a flat, relatively rock-free area almost on the equator that the engineers and managers on the projects liked because it was perceived to be a very safe site to land. What made it *the* top candidate landing site, though, was that the science team liked it so much as well. Several years earlier, as part of the *Mars Global Surveyor* orbiter's Mars-wide mineral-mapping campaign, Phil Christensen's TES team had discovered evidence for large deposits of a coarse-grained form of the mineral hematite (Fe_2O_3) in Meridiani Planum. These turned out to be the largest deposits of the stuff on the planet. Planetary scientists had known for years that there is hematite on Mars—but that was the fine-grained, pigmentary, "red" variety of hematite that is often found in oxidized soils on the Earth and that is thought to be at least partly responsible for the brownish-red color of the soils and dust on Mars. The coarse-grained, "gray" hematite discovered by TES is a different beast altogether. On Earth, hematite crystals usually grow to such large sizes only in the presence of liquid water. So, while models of how the fine-grained, red hematite formed on Mars might be just as likely to involve the presence of liquid water or not, it seemed much more likely that liquid water might be tied up in the origin of the gray hematite. The hematite deposit in Meridiani was like a big mineralogic beacon saying, "Looking for water? Land a rover here!"

Another of the finalists was a landing site inside a 160-km-wide impact crater named Gusev, about 15° south of the equator within some of the planet's ancient highlands terrain and almost exactly on the other side of the planet from Meridiani. Gusev has a 500-km-long, putatively water-carved channel called Ma'adim Valles flowing into it from the south, and so it was hypothesized that Gusev may once have been a crater lake, filled to some level with liquid water. Any lakebed sediments that may have been deposited on the floor of Gusev would be excellent candidates for preserving evidence that the water was there and that the environment may have been habitable. Our landing ellipse could fit easily inside the floor of Connecticut-sized Gusev Crater, but there was some concern that winds inside the crater might be too strong and pose the same kind of risks to the landing system as the modeled winds in Melas. The other concern was that we knew that Gusev was a "dusty" site, and so there was some possibility that we might spend most of our time studying the dust-covered outsides of whatever deposits we found rather than the deposits themselves. Some scientists were also worried that there might be a lot of volcanic rocks on the floor of Gusev covering up any putative lakebed deposits. While it seemed to many on the team that Gusev held a big geologic beacon saying, "Looking for water? Land a rover here," there was still a lot of uncertainty about both the scientific potential and the landing safety of the site.

The third possible site, Isidis Basin, was within a chunk of ancient southern highlands terrain that extends a little north of the equator. The proposed landing ellipse was in a relatively flat plains area almost exactly on the equator. Isidis was similar in some ways to Gusev because there was some evidence in the orbital images for possible surface water flows in the area in the form of valleys extending from the highlands into the plains where the landing ellipse would fit. There was a lot of controversy about the nature of those valleys, however, and no strong consensus that liquid water ever needed to be on the surface for significant amounts of time to form them. Isidis was one of the so-called "wind-safe" sites that the engineers and managers liked best, however, because the probability of landing safely was deemed to be high (around 90 percent, by some estimates). While the engineering folks liked Isidis a lot, the science community consistently ranked it below Meridiani and Gusev in terms of scientific potential.

The fourth finalist was the most discouraging one from a science perspective, because it was chosen almost entirely based on landing safety rather than any particularly compelling scientific interest. The project managers thought it would be prudent to keep on the table the most absolutely wind-safe place that they could find, just in case some new, late information came in about one of the other sites that would eliminate it from the competition (even after launch). A site was chosen in Elysium Planitia, on what was thought to be a flat and relatively rock-free volcanic plain, where all the models predicted a nearly 95 percent probability of a safe landing. It was never clear to me how much value to place in such estimates, because the only way to validate a model of the probability of landing safely would be to test it by trying to land safely in many different places. Now that would be a fun set of tests. I suppose that choosing a safe site with little or no consideration of the science goals of the mission could be considered a valid, conservative thing to do, from an engineering standpoint, but it was hard to swallow. There was just nothing thought to be related to possible past water or habitability at Elysium.

Ultimately, Meridiani and Gusev, ranked #1 and #2 by the scientific community, were selected by NASA Headquarters officials as the final landing sites. Meridiani had been a shoo-in from the beginning, so that was no surprise. Gusev had been a nail-biter because of the concerns about the winds. The engineering and science teams worked hard to convince NASA officials that there was still a very high probability of landing safely in Gusev, and that the higher scientific interest in landing there instead of in Isidis or (especially) Elysium meant that a safe landing in Gusev had much more scientific "payoff" potential than one at either of the other sites. It must have been a difficult decision for the upper-level NASA folks to make, intentionally choosing a more risky landing environment for the highest-visibility NASA project since the tragic loss of the space shuttle *Columbia* in 2003 and the unfortunate losses of both NASA Mars missions in 1999.

Up, Up, and Away

Though we now knew that we would land *Spirit* in Gusev and *Opportunity* in Meridiani, it would have been a pinnacle of hubris to think that we understood before we got there what these places would really be like. We had an idea of what the topography would be like, as well as some of the physical properties of the surface, based on orbital and telescopic remote sensing data. Would we really see evidence of sediments or water-carved features preserved in the geology? Would we see the kinds of marker minerals that are diagnostic of long-standing water-rich environments? Would there be other evidence to support or refute the concept of a warm and wet Mars—the kind of place where biologic processes could have possibly occurred (or—it's a long shot—could still be occurring)? Could we confirm on the ground what we think we see at these places from orbiting spacecraft? That is, could we provide the ground truth needed to back up all of the other global reconnaissance missions?

All of this potential, all these unanswered science questions, the endless meetings and viewgraph presentations, the long hours away from home and family, all of this effort expended by hundreds and hundreds of people from labs and universities and companies around the world in the name of science and exploration, are what made me simultaneously want to dance and vomit as I watched *Spirit* lift off from Cape Canaveral on that sunny day back in the summer of 2003. As the rocket cleared the tower and we jumped and waved our arms and cried with joy, that's when I realized that what was really inside that rocket wasn't just a machine, wasn't just the cameras I was responsible for. It was carrying all of us and our future to Mars.

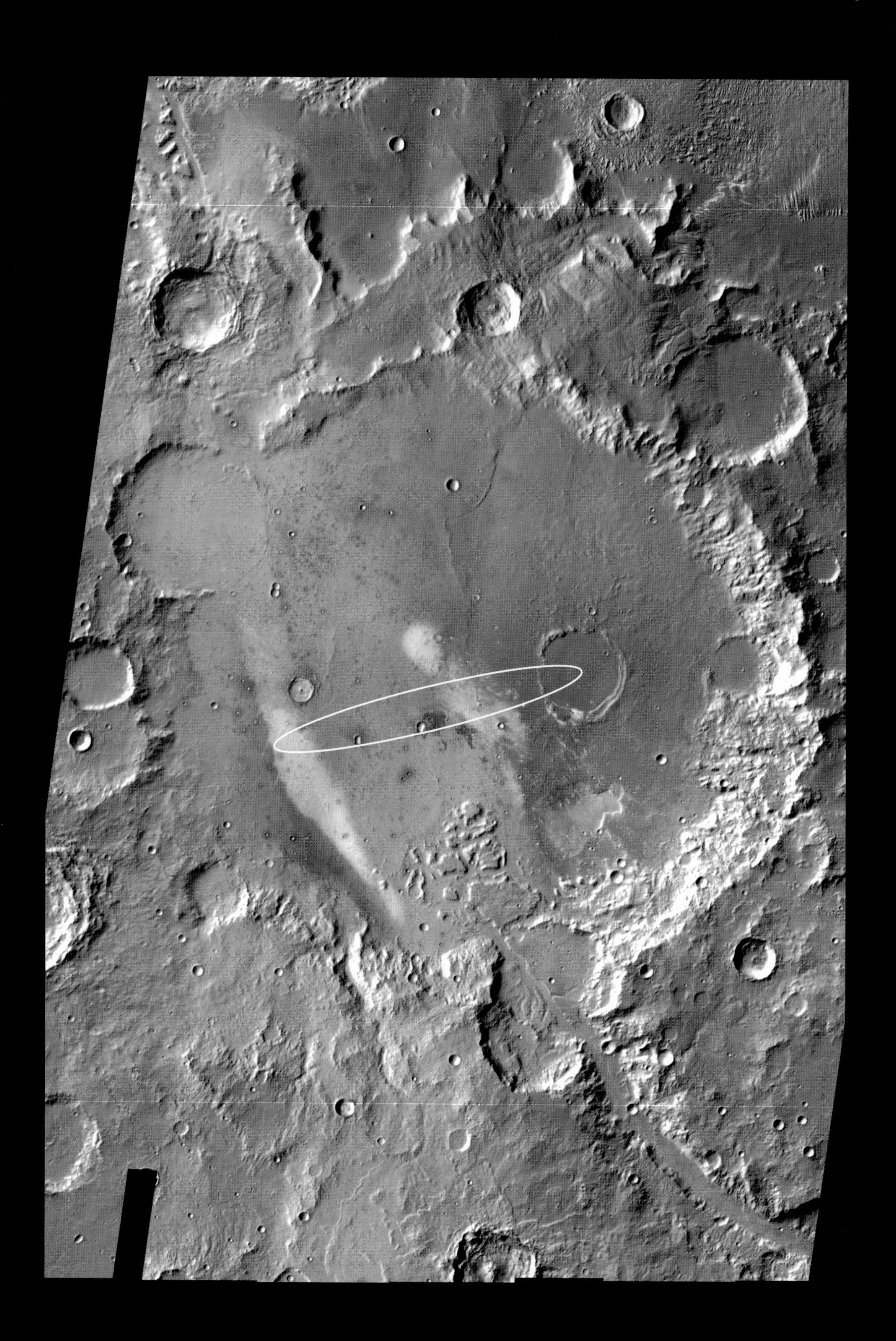

Part 2:
With *Spirit* in Gusev Crater

It's not every day that you land on Mars. In fact, before the rovers arrived, landing on Mars had been successful only three times out of ten attempts by NASA and the Russians over the past four decades. Years of stress and anxiety from building the rovers, testing them, fixing their problems, launching them, and guiding them on their long interplanetary journey were all boiled down to "six minutes of hell" in the words of one NASA official as they plunged into the martian atmosphere and prepared to land.

Cruising

The *Spirit* and *Opportunity* launches went off without a hitch, thanks to the experience and expertise of the U.S. Air Force/Boeing Delta rocket crew and the support staff from NASA's Kennedy Space Center and JPL. Except for an occasional minor course correction and thruster firing on the way and a few modest instrument checkouts, the cruise to Mars is normally a quiet and relaxed time for the spacecraft. Between launch and landing, Isaac Newton and his famous laws of physics were really in control of the mission. Back on Earth, it was anything but quiet and relaxing. We struggled to finish the software. Sometimes we bumbled through grueling real-time mission simulations designed to find out where humans, hardware, and software might break down. Our logic was that we'd much rather find as many flaws and bugs as we could during cruise, to minimize the number we'd probably discover once we got to Mars.

Things weren't really that quiet and relaxing on the spacecraft, either. While we were flying through interplanetary space the rovers were bombarded by high-energy particles coming from some of the biggest solar flares in recorded history. These amazing outbursts of energy from the Sun produced some spectacular and beautiful nighttime displays of the northern lights all across the world (even in upstate New York, which is usually too far south to get good views of the aurora). They also represented a real potential danger to the rovers. High-energy particles can cause havoc with sensitive electronics like detectors and computer memory chips. This is why some power and communications systems on Earth and satellites in Earth orbit are sometimes damaged after strong

Gusev Crater and the *Spirit* landing ellipse, superimposed on a colorized *Mars Odyssey* THEMIS infrared image of the crater. (NASA/JPL/ASU/J. BELL)

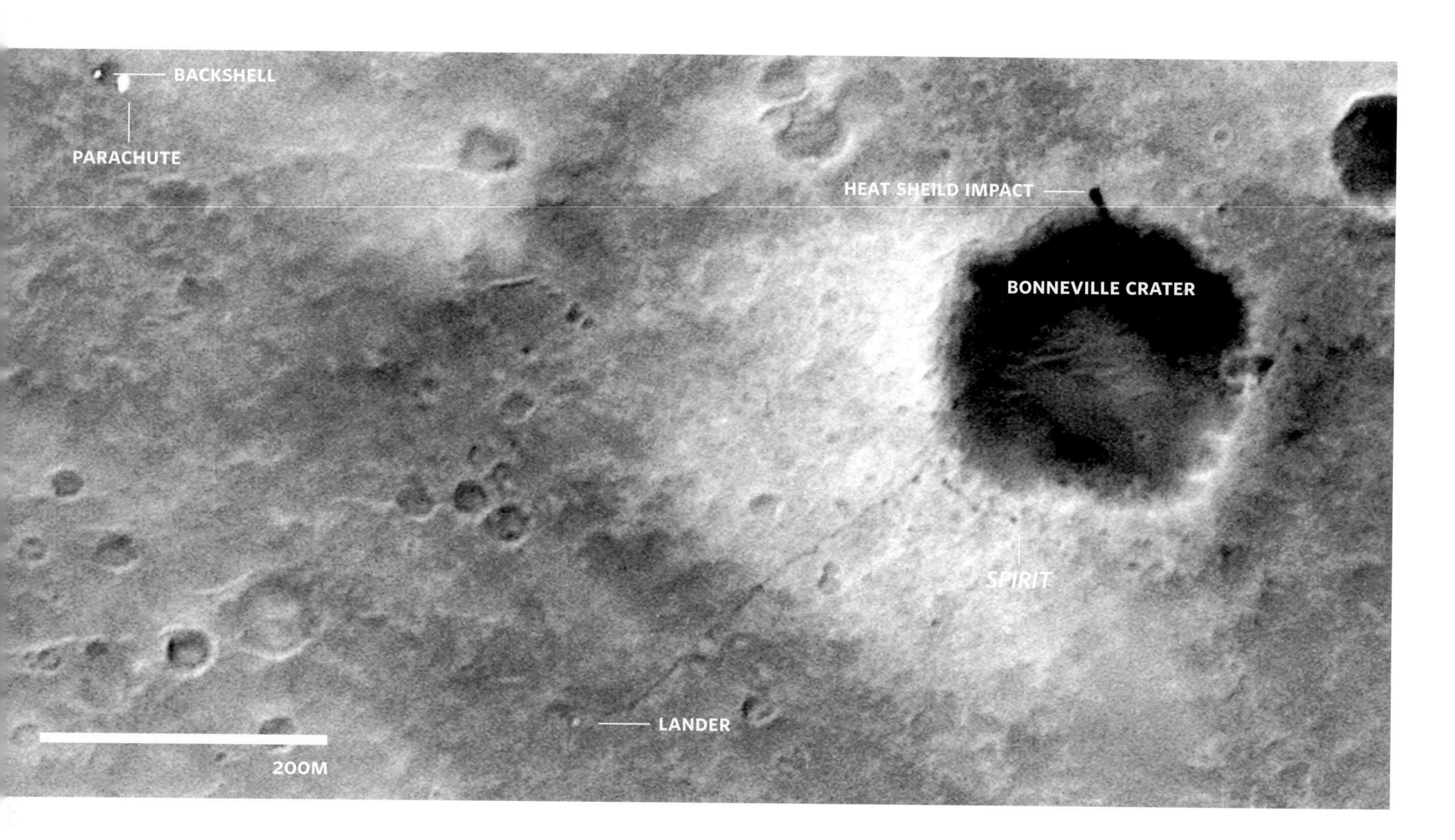

LEFT TO RIGHT
Mars Global Surveyor Mars Orbiter Camera image of *Spirit* at her landing site in Gusev. The rover was easily spotted from orbit by following her tracks up to Bonneville Crater. (NASA/JPL/MSSS)

Orbital view of *Spirit*'s landing site in Gusev Crater, including the nearby Columbia Hills. (NASA/JPL/CORNELL)

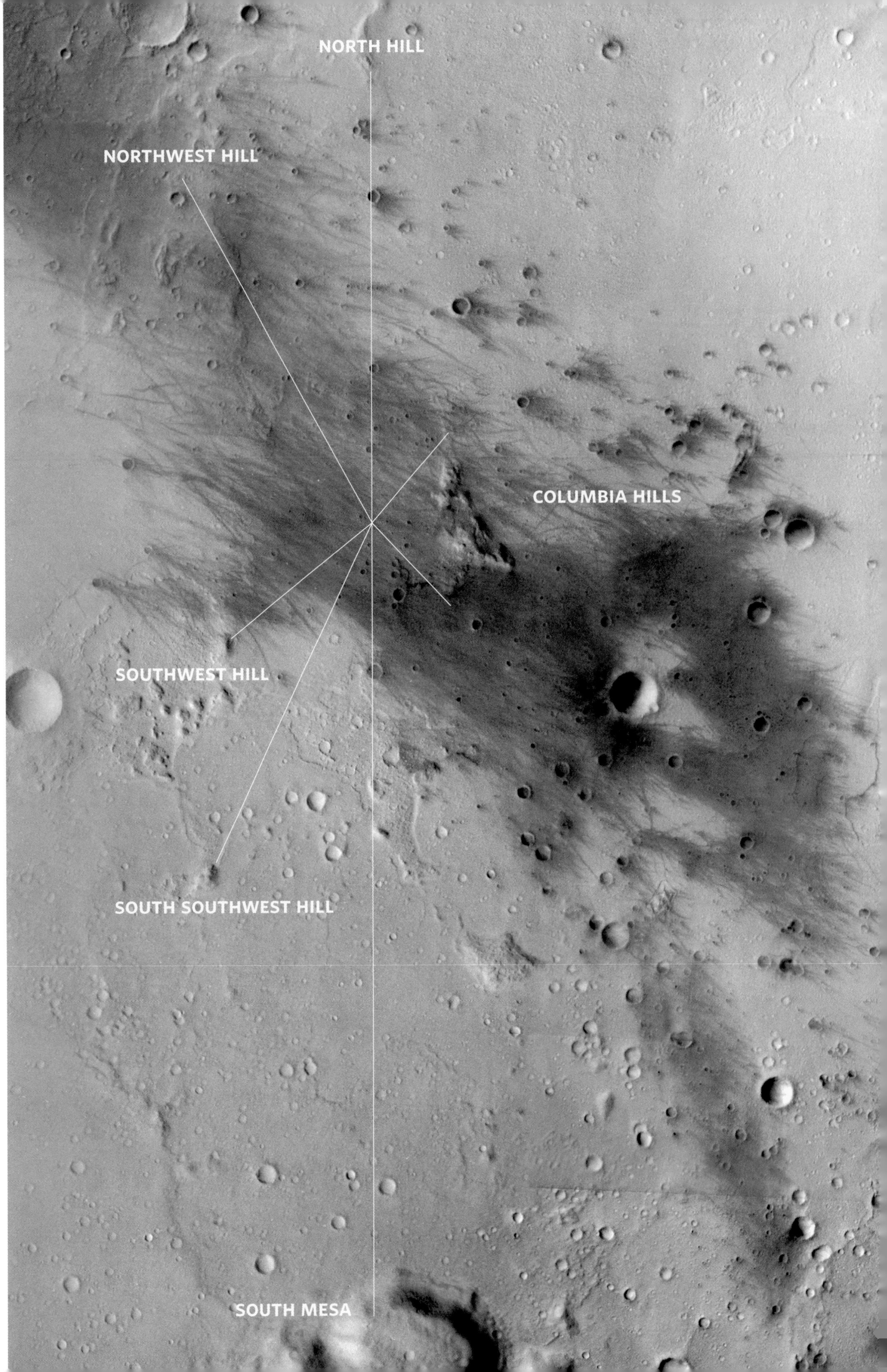

NORTH HILL

NORTHWEST HILL

COLUMBIA HILLS

SOUTHWEST HILL

SOUTH SOUTHWEST HILL

SOUTH MESA

LEFT TO RIGHT
Computer simulations of the rovers' postlanding deployment sequence. (NASA/JPL/MAAS DIGITAL)

FOLLOWING SPREAD
Spirit's first postcard home, from the Gusev landing site on SOL 2. (NASA/JPL/CORNELL)

solar flares. Of course, we had an abundance of these kinds of electronics on both spacecraft. We monitored the health of the rovers more carefully and more frequently, especially after the biggest particle "storms" swept past. Tests showed that we had received our fair share of battering by these particles, but nothing was damaged. There was one part of the computer's memory, however, that could not be tested, and it was an absolutely critical part for the landing events to work right. So, just in case, the engineers reinstalled the software into that part of memory during the later part of our cruise, just in case something had gone wrong.

Landing *Spirit* and *Opportunity* on Mars was not easy. Each spacecraft comes screaming in from interplanetary space at more than 5 km/sec (more than 12,000 mph). It has to hit the atmosphere at a precise angle to allow friction to slow it down but not burn it up. Next, it has to perform a Rube Goldberg-like series of explosive latch releases, parachute-and-airbag deployments, and other on-board computer-processing tasks. All of these events are crammed into the most important few moments of the entire mission. If all went well, less than an hour after first contact with the upper atmosphere of the planet, each rover should be safely on the ground and out of its cocoon. There are thousands of things that could go wrong, but the system was tested in numerous simulations and real airbag drops. The engineers felt as confident as they could that they had tested and built in redundancy as much as possible, given the time and money available. It was truly as robust a landing system as it could be, given that the only true tests of the system were the ones that happened on Mars in January 2004. Still, no one can truly have a warm and fuzzy feeling about rolling those dice.

Like *Mars Pathfinder*, the *Spirit* and *Opportunity* rovers were carried to Mars inside a tetrahedral lander structure that unfolds, flower petal-like, after landing. Unlike *Pathfinder*, though, these landers do not carry cameras or power supplies or communications equipment, but are instead simply

Spirit "Mission Success" panorama, SOLS 3–5. (NASA/JPL/CORNELL)

"pallets" with rovers on top. After bouncing, landing, and unfolding, each lander+rover would be sitting atop a pile of deflated airbags that themselves might be on top of rocks or dunes, possibly tilting the whole structure. The scary part was that we would then have to tell the rover to disconnect from the lander and somehow drive off onto the ground. We didn't know exactly which way we'd drive off, because that depends on what obstacles are around us. We didn't know exactly how far we'd have to drop off the lander down to the ground, because that depends on the state of the airbags and the tilt of the lander. We didn't really know how long it would take to get off the lander. It would be a methodical, conservative process that includes a number of real-time "go/no-go" decisions among flight engineers back on Earth. In some scenarios it might only take four or five martian days (sols) to get the rover off the lander, whereas in other scenarios it might take six or seven sols or longer. We knew we'd spend part of this time looking around and taking some big, panoramic mosaics to try to get the feel for the place before heading out. However, the mission wouldn't really start until we had those six wheels in the dirt.

Spirit landed on January 4, 2004. Some people were cheerily optimistic beforehand; others walked around muttering to themselves about "one more test" or "I hope the redundancy kicks in." I felt the same sense of fatalism that many of the engineers were expressing—we'd done everything that we could within the limits of time and money and technologic skills to ensure success, and now it was really up to *Spirit* and Mars. A hundred million miles away, a computer was executing a sequence of commands that the team had programmed and tested a year earlier. In fact, because at the time of landing, radio signals from the rover were taking nearly ten minutes to get from Mars to Earth at the speed of light, the whole thing was actually over before we could even know what had happened. Fatalism seemed appropriate. We watched, and paced, and waited.

Spirit's SOL 9 postcard looking out across Sleepy Hollow and the bounce marks left by the lander. (NASA/JPL/CORNELL)

FOLLOWING SPREAD
A beautiful postcard from *Spirit*, acquired on SOL 5. (NASA/JPL/CORNELL/DON DAVIS)

The whole spacecraft was dropped to the surface to bounce around like a beach ball. The first bounce hit the ground at about 25 meters per second (60 mph), and the rover ultimately bounced about a dozen times before finally coming to rest. The airbags deflated, the lander's petals opened to expose the rover inside, and the rover unfolded its solar panels, deployed its mast, and took its first look around.

We saw all of these events occur not in pictures but only through simple computer plots and tables of numbers on our monitors, much like the stereotypical images of the controllers at mission control in Houston monitoring the astronauts in space. Of course, no one was there on Mars to witness this extraordinary sequence of events. I couldn't help but wonder what a hypothetical martian out for an afternoon stroll would have thought of this fireball streaking through the sky, coming to a fiery stop just above his head, and crashing onto the ground with an almost comical bounce, bounce, and roll. That this was a visit from the blue planet next door might not be my first guess.

Just after the rover phoned home to tell us that it had survived, I caught a telling glance from Rob Manning, one of the JPL rover engineers who had orchestrated the landing. "We did our part," his glance told me, "now it's time for the scientists to come through." Did the cameras survive undamaged? For the first time, I felt the full weight of the project fall onto my shoulders and those of my science team colleagues. Now *I* was walking around muttering about lens caps and light levels, craving a coffee, and wondering if we were really ready. The same pit in my stomach that

LEFT TO RIGHT
An evocative, emotional postcard of the *Spirit* lander at Gusev. (NASA/JPL/CORNELL)

Finally driving away from the *Spirit* lander, SOL 39. (NASA/JPL/CORNELL)

FOLLOWING SPREADS
52–53
Pancam panorama looking out toward the Tennessee Valley, beyond Husband Hill, *Spirit* SOL 510. (NASA/JPL/CORNELL)

54–55
The "Everest" panorama, taken by *Spirit* from the summit of Husband Hill on SOLS 620–622. The Marsdial —dusty but still working well—can be seen on the rover's rear solar panel. (NASA/JPL/CORNELL)

56–57
Spirit shot this 360° false-color (for mineral analysis) Pancam view, called the "Seminole" panorama, on SOLS 672-677 while descending the southern slopes of Husband Hill. (NASA/JPL/CORNELL)

I had last felt at launch seven months earlier was back. Within a few hours we'd either see the first pictures and other data from the rovers, or we'd have to start an agonizing investigation of whatever was wrong.

Photos Through Space

Sending photographs back from outer space is complicated. There is a small armada of spacecraft wandering around the solar system, and NASA keeps up with them using a series of large radio telescopes spaced around the Earth. There are outposts in Goldstone, California; Madrid, Spain; and Canberra, Australia. This is the Deep Space Network, or DSN, and it operates twenty-four hours a day, seven days a week for two-way communications with the armada. Time on the DSN is precious and carefully negotiated, especially on the most sensitive 70-meter-diameter radio telescopes, because so many spacecraft require commanding and monitoring and some of them need to be in continuous contact with Earth during critical maneuvers, flybys, or landings.

Typically, spacecraft will acquire their pictures and other data and then encode them into radio signals. They transmit these weak signals back to Earth, where they can be detected and decoded by the DSN antennas. The rovers are capable of this kind of direct-to-Earth communication, but with two important constraints. First, the Earth must be visible to the rover's transmitting antenna. This

LEFT TO RIGHT

Close-up Pancam-colorized Microscopic Imager view of the 4.5-cm-diameter RAT hole ground into Humphrey, *Spirit* SOL 60. (NASA/JPL/USGS/CORNELL)

"Mimi," the first example of a possibly layered rock found in the Gusev plains. (NASA/JPL/CORNELL)

Martian light and shadow near the rock "Sandia," *Spirit* SOL 53. (NASA/JPL/CORNELL)

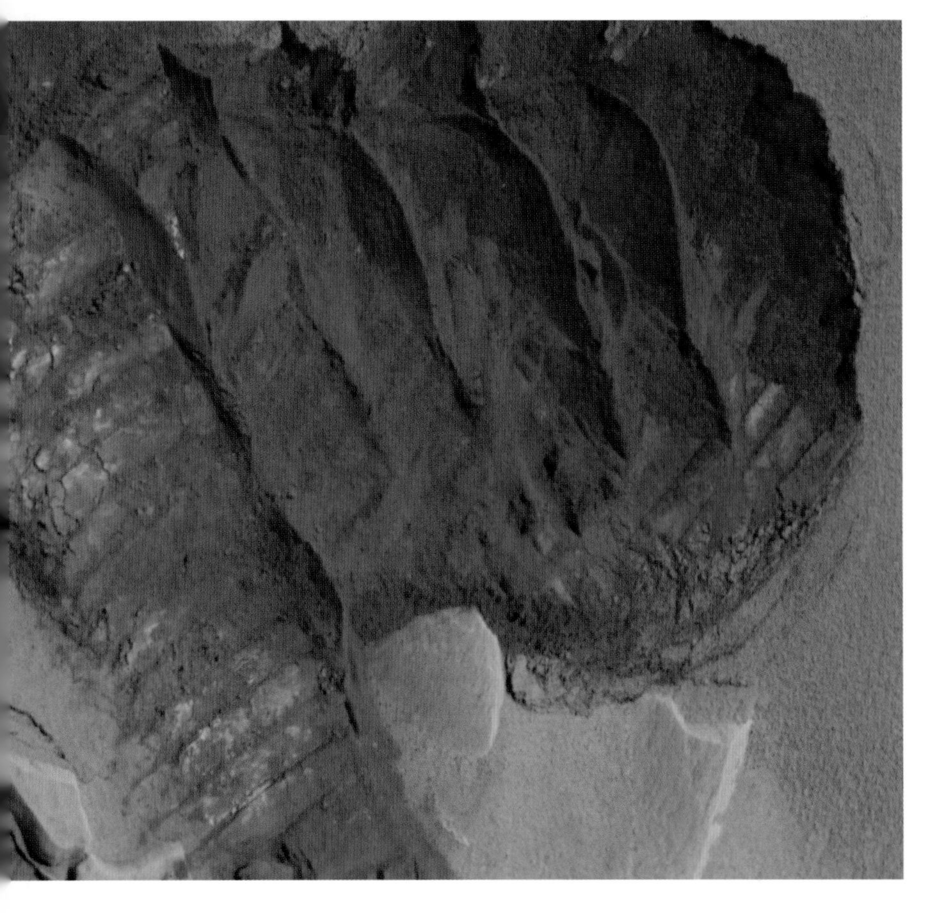

LEFT TO RIGHT
SOL 72 image of a 15-cm-wide *Spirit* wheel scuff mark in a small sand dune. (NASA/JPL/CORNELL)

Pancam-colorized Microscopic Imager view of dust, sand, and millimeter-size pebbles in the walls of a trench dug into Serpent dune, *Spirit* SOL 70. (NASA/JPL/USGS/CORNELL)

Pancam-colorized Microscopic Imager view of the 4.5-cm-wide RAT hole ground into "Mazatzal," *Spirit* SOL 85. (NASA/JPL/USGS/CORNELL)

isn't always the case, though, because Mars spins and the Earth is only "up" for about twelve hours out of each twenty-four hour, thirty-nine minute Mars day or sol. For half of each sol, then, direct-to-Earth communications aren't possible. Second, communicating this way requires a lot of power. The rover has to try to send the strongest signal possible to Earth, because the signal dissipates along the way and must remain strong enough to be picked up by the DSN antennas. Power spent by the rovers transmitting data means less power for driving, taking pictures, or other operations. Like everything else, communication time is a resource that has to be carefully managed.

All during *Spirit*'s entry, descent, and landing phase on the evening of January 4, 2004, the Earth was in constant view of the spacecraft, which is why we could receive the direct-to-Earth telemetry that told us that things were going well. However, shortly after the landing and before the rover finished unfurling from the lander, the Earth set below the horizon. This made it impossible to diagnose the rover or to get the first pictures and other data back by direct communications to Earth.

To prevent such a blackout period from occurring, the team attempted a bold experiment. It was known that if all went well, just after the rover finished unfurling and acquiring its first pictures, the *Mars Odyssey* spacecraft, a NASA orbiter some 400 km overhead and studying Mars since 2002, would pass right over the *Spirit* landing site and be visible above the horizon for about ten minutes. The rover had been preprogrammed to know what time *Odyssey* would pass over. At that time it would attempt to uplink the images and other data directly to the orbiter rather than to the Earth. *Odyssey* has a special receiver cleverly designed in advance for just such an opportunity. If *Odyssey*

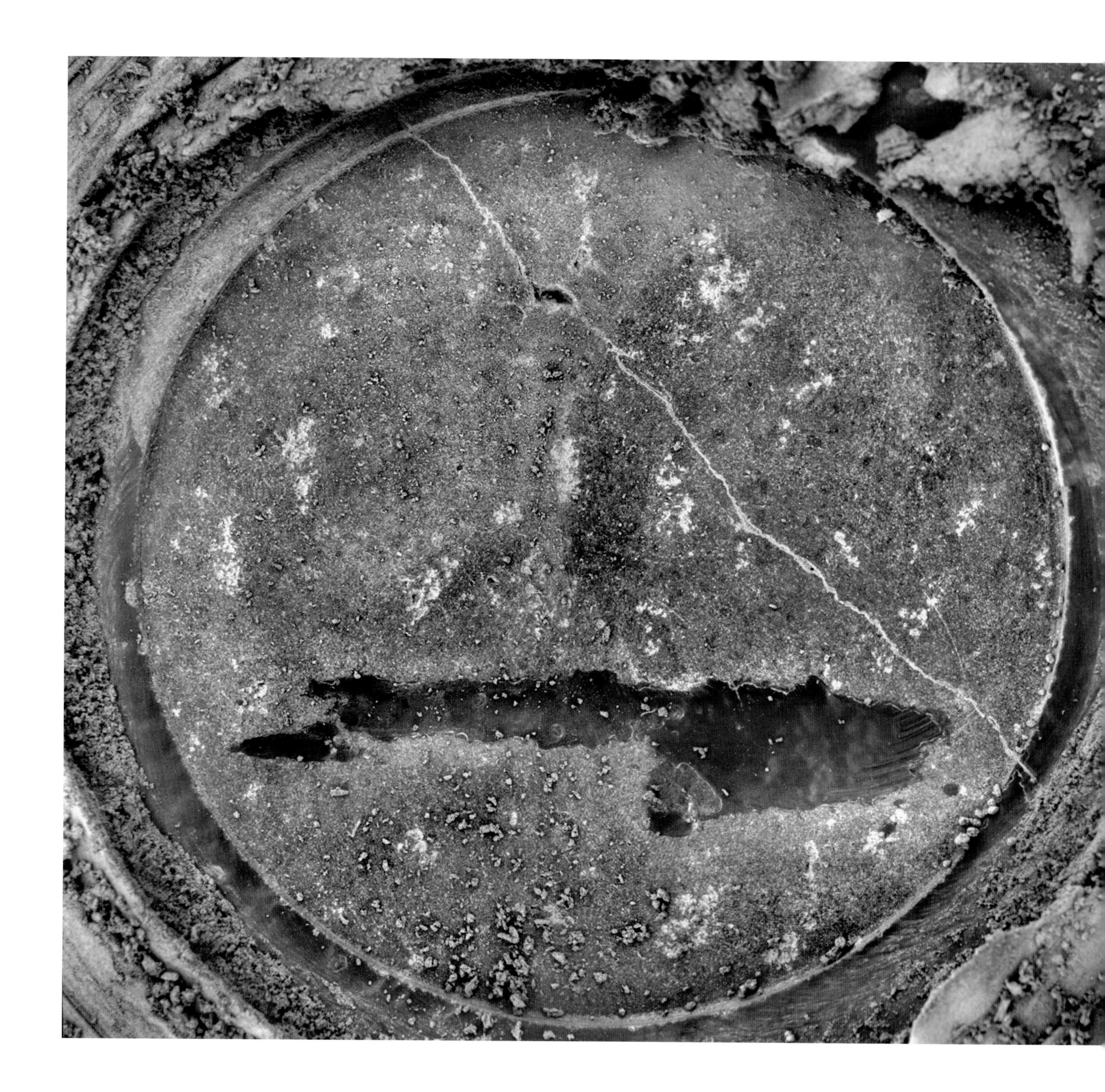

received the data, it would relay the images and other information back to Earth, where they would be picked up by the DSN. This was essentially the first attempted satellite phone call from another planet, and it marked an important first step in NASA's plans to use "local assets" like orbiters as part of a Mars communications infrastructure.

If this experiment worked, it would lead to two significant advantages for the mission. First, we could double or triple the amount of data that we could send back compared to the direct-to-Earth method because we could communicate with the rover after the Earth had set. A higher volume of data could be sent to and stored on *Odyssey* because the orbiter is much closer to the rover, which means the radio signal is much "cleaner." This was near and dear to my heart because if it worked, it would especially benefit the Pancam cameras, which would be taking all of the color and high-resolution panoramic data during the missions. Second, it would require much less power for the rover to transmit signals to *Odyssey* than to Earth, and that power savings could be spent on taking more pictures or other science data or driving. In addition, if we could get the experiment to also work with NASA's other orbiter, *Mars Global Surveyor,* we could stand to gain even more.

Even though the Earth had set in Gusev Crater, we knew that if the *Odyssey* relay experiment worked, we'd be getting our first *Spirit* images at the DSN a few hours after landing. On cue, *Odyssey* passed over and *Spirit* phoned home. Once picked up by the DSN, the data was routed to JPL, where it was converted into the digital pictures and other information we'd been waiting for. Our laptops and workstations suddenly came to life with the first views of Gusev Crater seen by humans.

That moment was thrilling on so many levels. As we gaped, we saw a ridge off on the horizon, some intriguing rocks and pebbles to study nearby, and some enigmatic bright and dark soil patches that looked like they might have been scraped by the airbags. Mostly, though, I was simply delighted and relieved that the cameras worked! The team was finally ready to take the photographs that we'd been dreaming about for years. We stared in wonder at those first few images, like everyone else in the world. Then we jumped and yelled and wept and laughed and danced like fools for a while, relishing what will certainly be one of the highlights of our scientific careers.

Once we knew that we had fully functional cameras, we had to devise a strategy for taking as many photographs and sending ("downlinking") back to Earth as many as we could on every SOL on Mars. In devising the strategy we had to make a number of compromises. Pancam is what would be

Spirit's "Cahokia" panorama, revealing the picturesque rocks, soils, and outcrops of the West Spur of the Columbia Hills. It took 11 SOLS (213–223) to acquire the 470 images needed for this view.

called a "one-megapixel" digital camera at your local home electronics store. With two of them on each rover (for stereo vision and redundancy), each pair of images requires more than two million picture elements or pixels to be downlinked. If we radioed these pixels back at their full fidelity and resolution, it would require about 25 megabits (25 million ones and zeros strung together) of data. However, we were typically able to downlink only about 50 to 100 megabits of data per SOL from each rover, so we clearly didn't want to use up a quarter to half of all the data with only a single pair of images! Also, if we kept taking full-quality images like these, we'd quickly fill up the rover's onboard memory, which would prevent us from doing anything else until we could downlink those images and free up the memory.

We had to find a way to compress the images to squeeze more of them into the downlink without losing too much quality. It's the same problem digital camera manufacturers have to tackle: The number of images that you can store on your flash memory card depends on how much you're willing to compress them and give up on some of the image quality. In our case, we had several methods of compressing the images within the rover's computer before they were sent back to Earth. Sometimes we shrank them from 1024 x 1024 pixel images down to 512 x 512 or to 256 x 256 size, trading resolution for more pictures. Most of the time, though, we used what image processing people call "lossy" compression. This uses a software program on the rover's computer that hunts for the parts of the images that are bland—for example, in areas like the sky, or sand dunes, or parts of the rover deck. The quality of the images can be degraded in these areas without losing much information. In other regions, where there were interesting, small-scale geologic features, like on the rocks or in the distant hills, we would preserve the quality. We tested this program before launch and found that we could decrease the quality of an image by as much as a factor of ten this way without a dramatic difference detectable to the eye. Image compression was a critical part of our rover toolbox; without it we would not have been able to downlink most of the spectacular images that we obtained.

Another important compromise had to be made about colors. Each Pancam camera has a small, eight-position filter wheel on the front that allows us to obtain images in different parts of the spectrum that we can combine into color images. For example, to take a typical color image, we would acquire three separate images, one each through red, green, and blue filters. We would downlink all three, and then combine them into a color composite using computers back on Earth. This is essentially the same thing that happens inside any color digital camera that you can buy off the shelf today, except that those cameras have fixed filters mounted inside and the processing and combining of the three color images into a color composite all happens automatically inside the camera's computer chip after you snap the shutter.

I've often been asked why we didn't just send something like a commercial, many-megapixel, off-the-shelf color digital camera to Mars instead of our more complex and expensive one-megapixel, filter-wheel-based color camera system. There are a couple of important reasons. First, is the environment. The atmosphere of Mars is so thin that it's almost a vacuum. It also gets incredibly cold, down to -100°C at night and only up to about 5°C at its warmest during the day at the landing sites. An off-the-shelf digital camera in those conditions won't last long. Second, the light-sensitive pixel detectors inside most digital cameras are capable of seeing a much wider range of colors than the human eye can, from the ultraviolet through the infrared. We wanted to be able to exploit that capability (usually ignored in commercial digital cameras) because certain kinds of rocks and minerals absorb and reflect these "extreme" colors in special ways. Another advantage of going with a filter-wheel system like this was that we could put opaque, welders' glass–style filters in two of the slots so that we could look directly at the Sun with the cameras. This allowed us to use the cameras

The first three color images returned by *Spirit* from Mars, just a few hours after landing. The Pancams work! (NASA/JPL/CORNELL)

as a sort of high-tech sextant. We could spot the Sun as a way to get our bearings, monitor how much dust was in the atmosphere every sol, and also watch the martian moons Phobos and Deimos pass in front of the Sun during occasional martian eclipses.

A third reason why we couldn't send an off-the-shelf digital camera to Mars is less intuitive. NASA has a policy about flying instruments in space that boils down to "It can't fly in space unless it's already flown in space." That's a bit of a stretch, of course, because if it were true, NASA could never fly anything in space. But it's not far off the mark. Space instruments and systems are judged partially on a factor known as their "heritage"—that is, have they flown in space successfully before, or are they a simple descendant or relative of something else that has already flown? This means that cameras or other instruments that are the latest high-tech gadgets or that rely on some brand-new technology don't yet have spaceflight heritage. Thus, they will be judged to be inherently more risky than other perhaps more mundane or simple, but proven, instruments. We can't yet fly, for example, a 10-megapixel, HDTV-compliant, 20-GByte flash memory digital camera in space. Nothing like it has ever flown in space before. The best we could manage to convince others that we could build and fly and operate at the time we were advocating our design was our one-megapixel camera. It was only slightly more advanced and risky compared to digital cameras that had been successfully used in deep space before, and so its heritage was judged to be high.

These kinds of choices meant that we had to design a custom camera system capable of surviving in an extreme environment. We also had to endow it with custom filters that would allow us to try to detect the most interesting kinds of surface materials. A similar set of constraints had guided the design of the *Mars Pathfinder* mission camera, which acquired stunning images from the surface of Mars back in 1997. Because I was fortunate enough to have been a member of the camera team for *Pathfinder*, I was able to help combine the best aspects of that design with the requirements of our new rover mission to bring Pancam to life.

The problem with all those filter choices, though, is that we just didn't have enough downlink bandwidth to take pictures of every scene using all of them. So we had to prioritize and compromise. Often times we had to scale back. Among the highest priority filters were the ones that allowed us to make "true color" composites of the rocks, soils, and other features at the landing sites. Using some image-processing software and images through the red, green, and blue Pancam filters that are closest to the behavior of the red, green, and blue receptors in the human eye, we could make our best estimates of what the view would look like if we were standing there ourselves. Getting the colors right from cameras in space is a challenge, but our MarsDial helped convince us that we were doing it right.

Postcards

Steve Squyres and I made an unusual decision early on in this project. If we were lucky enough to land a rover (or two) on Mars safely, we decided that we would share the images with everyone without restrictions or embargoes, as quickly as possible, using the Internet. Both of us had been involved in previous projects where this had not been the case. Certainly there were technological limitations on data distribution in the days before the Internet. Sometimes, though, scientists who have put so many years of their career into a project feel entitled to "own" the data. It's natural for some people to feel a close personal connection to images or other data that have been acquired at a cost of years of hard work and personal sacrifices. However, sometimes people can go too far and start referring to the data as "my images" or "my spectra." In the long run, I believe that holding the

A postcard from the *Spirit* "Legacy" panorama, showing dark tracks in the light soils halfway to the rim of Bonneville Crater, SOLS 59–61. (NASA/JPL/CORNELL)

FOLLOWING SPREAD
A triptych of the *Spirit* Santa Anita panorama, in the plains near the base of the Columbia Hills, SOLS 136–141. (NASA/JPL/CORNELL)

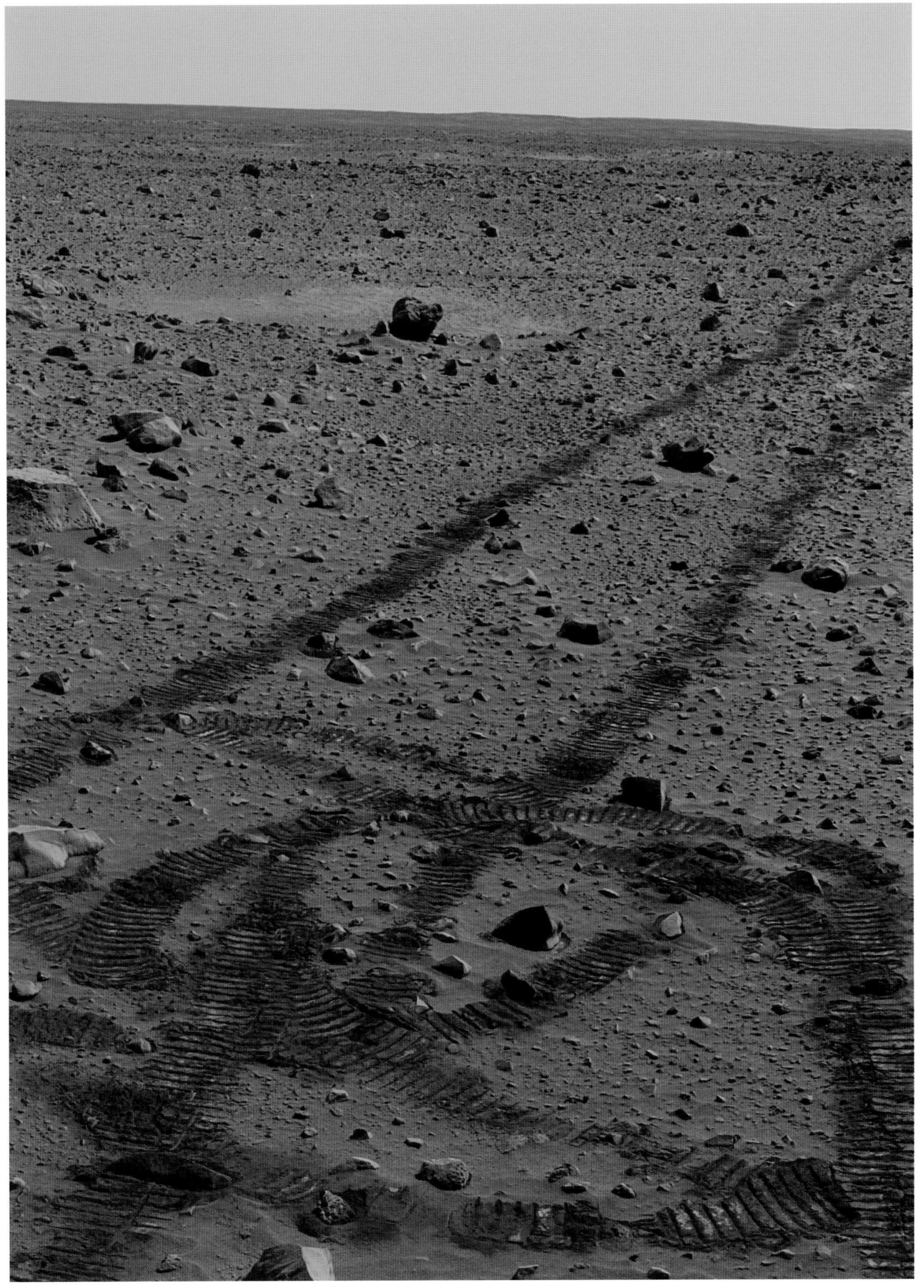

cards too close to your chest like this can do more damage than good for science and exploration. We were privileged to have been entrusted with taxpayer dollars to run this mission. We have an obligation to share both our successes and failures openly and honestly with the general public. In the case of the rovers, when the images are decoded at JPL from the radio signals the rovers send from Mars, a computer program automatically generates a JPEG version of every image at the same time, and these get posted on a publicly accessible Web site (see "Additional Resources" at the end of the book) usually within a day of being taken on Mars. As best we can count, millions of people have been accessing and downloading these images from the Web. Some of our colleagues think we're fools to have done this ("You're giving *your* data away!"). It is true that we have been "scooped" a few times on scientific papers or media stories by some people who use these instant images to get a quick result published. To us, though, that's a small price to pay to allow so many others—kids, teachers, space enthusiasts, laypeople, even members of Congress—to be able to follow along in near-real-time and to be a part of our amazing martian adventure.

I do a lot of traveling, both as part of my job and for vacation fun. One of my favorite things to do when I go to a new place is to visit the gift shop in the local airport or in the downtown mom-and-pop stores and check out their collection of postcards. Postcards are windows into the soul of the locals. Sometimes they are focused on the usual attractions: mountains or cathedrals, or the boardwalk by the ocean. Sometimes they sample the truly offbeat or just plain odd. We took a similar attitude with our first sets of images to be acquired by *Spirit*. I wanted them to be postcards—views showcasing the beauty of the natural environment that we now found ourselves in. We even set the first few mosaics up to be rectangular in shape, just like postcards. Of course, we didn't know before we got there what the place would look like, so designing the initial imaging sequences months in advance to take the pictures carried a sense of mystery, uncertainty, and wonder. Would we see sights that were mundane, or would some truly exotic or just plain odd wonders reveal themselves?

Spirit's first color postcard home was acquired the morning after the landing, and it covered a swath of reddish-brown ground extending from the airbags at the base of the lander out to the horizon. I had conflicting first reactions when the image popped up on my screen: At once I was elated because everything appeared to have worked! The sequence of thirty-six images (a mosaic three images wide by four images tall, in three filters) went off without a hitch. We used MarsDial images to calibrate the data, and the colors came out close to what I was expecting based on looking at Mars through a telescope with my own eyes. The mosaicking and analysis software built so painstakingly by Cornell team members Jonathan Joseph and Jascha Sohl-Dickstein stitched the images together like a charm. At the same time, though, I got this sinking feeling when I looked at what was actually *in* the image. It was flat, there were only a few small rocks, and there was nothing interesting in the distance. In short, it seemed boring. My conflicting emotions were resolved on the side of excitement, though, when I took a closer look at the image, which consists of nearly 13 million individual pixels. I loaded it into Jonathan's Pancam image-viewing program and zoomed in...more...more...more...until I was finally at the full resolution of the camera. At that scale, which is difficult to reproduce on the printed page, I could begin to see that the rocks are different from one another: Some are angular, some are pitted, and some are smooth and rounded. By wind? Or water, maybe? There are indeed faint hills and mesas off on the horizon, but they are partially obscured by dust and distance. There are several bright circular depressions—probably impact craters. Nearby, next to the airbags, I could see what looked like scratch marks in the soil that were created when the deflated airbags were retracted back into the lander by a set of cables. The scratches looked very strange, though, like someone had taken a carpenter's plane and dragged it across the ground, so that pieces of soil were lifted and curled up like wood shavings. A place that at first glance seems commonplace turns out to be quite alien after all.

Spirit's second postcard home, acquired a few sols later, is much more dramatic, both from a scientific and a photographic standpoint. The scene is looking back across the rover. It extends from the back of the solar panels past the lander and past a jutting-up piece of very dirty, partially deflated airbag material, and out to the horizon. The horizon appears tilted, as if we're on the side of a hill. This is an illusion, though: It is actually the rover that is tilted slightly on the lander deck. The hardware in the foreground gives the scene a sense of depth that is often difficult to convey in landscape photography. Zooming and scrolling around among the 23 million pixels in this postcard reveal many interesting geologic features. There are small sand dunes and ripples to the right, attesting to the active role of wind in continuing to shape the geology of places like this on Mars even today. There is also a curious bright-floored circular depression to the left that is relatively free of rocks. Depressions like this are seen elsewhere around the rover, too, and came to be called "hollows" by the team. They may be impact craters, perhaps so-called secondary craters formed when large blocks of rock dug up by an initial primary impact event later re-impacted the surface. This particular one, dubbed Sleepy Hollow, has marks on its floor that appear to have been created by the airbags as they came rolling and bouncing through on their way to the eventual landing site. Just below Sleepy Hollow, two medium-sized rocks can be seen, and the one on the right, which I named Sushi, appears to have a big hole through its center. How did it get like that? Why aren't there any rocks in the hollows? Are the dunes actively moving today? Is that reddish, lumpy thing sticking up behind the rover and blocking the ramp really a piece of dirty airbag, or is it a large rock?

Stitching It All Together

During the next week, while the engineers were busily working to prepare the rover for egress (driving off the lander), we scanned the cameras around to build up our first 360° color panorama of the landing site. This was dubbed the "Mission Success" panorama, because part of one definition of "success" for *Spirit* was to acquire and downlink this big opening color image. Acquiring color panoramas is a painstakingly slow process for the rovers, as the field of view of the Pancam is only 16° x 16°, or roughly equivalent to the view in a 35-mm camera with a 100-mm lens on it. There needs to be a small overlap from image to image, which means that it takes twenty-seven separate images to cover one 360° scan around the rover. Usually we have to stack at least three of these scans on top of one another so that we can cover the scene from just above the horizon down to the rocks and soils close to the rover. That's eighty-one images, then, just to obtain a panorama through a single filter. On average, it takes about ninety minutes to acquire that many images. However, we have to quadruple those numbers to obtain a panorama in three colors plus stereo vision. That kind of panorama (the minimum 360° color panorama that we would acquire) consists of 324 separate images and requires more than six *hours* (usually spread over several different sols) to collect. Add more time and more images if more color filters are needed, and you get a sense of how much time is spent and how much data is generated to make these photographs. They can take three to six sols to acquire, and sometimes a week or more to get their 300 to 500 million (or more) bits of data back to Earth.

For the initial *Spirit* panorama, we only had a few hours of imaging time per sol. The engineering activities to prepare the rover to drive off the lander had priority, so it took more than four sols to slowly acquire the pieces of the Mission Success pan. This added even more complexity to the

Earth is the pale dot in the center of this view of the predawn sky above Gusev Crater. This is a composite of *Spirit* SOL 63 Navcam wide-angle and Pancam high-resolution images.
(NASA/JPL/CORNELL/TEXAS A&M)

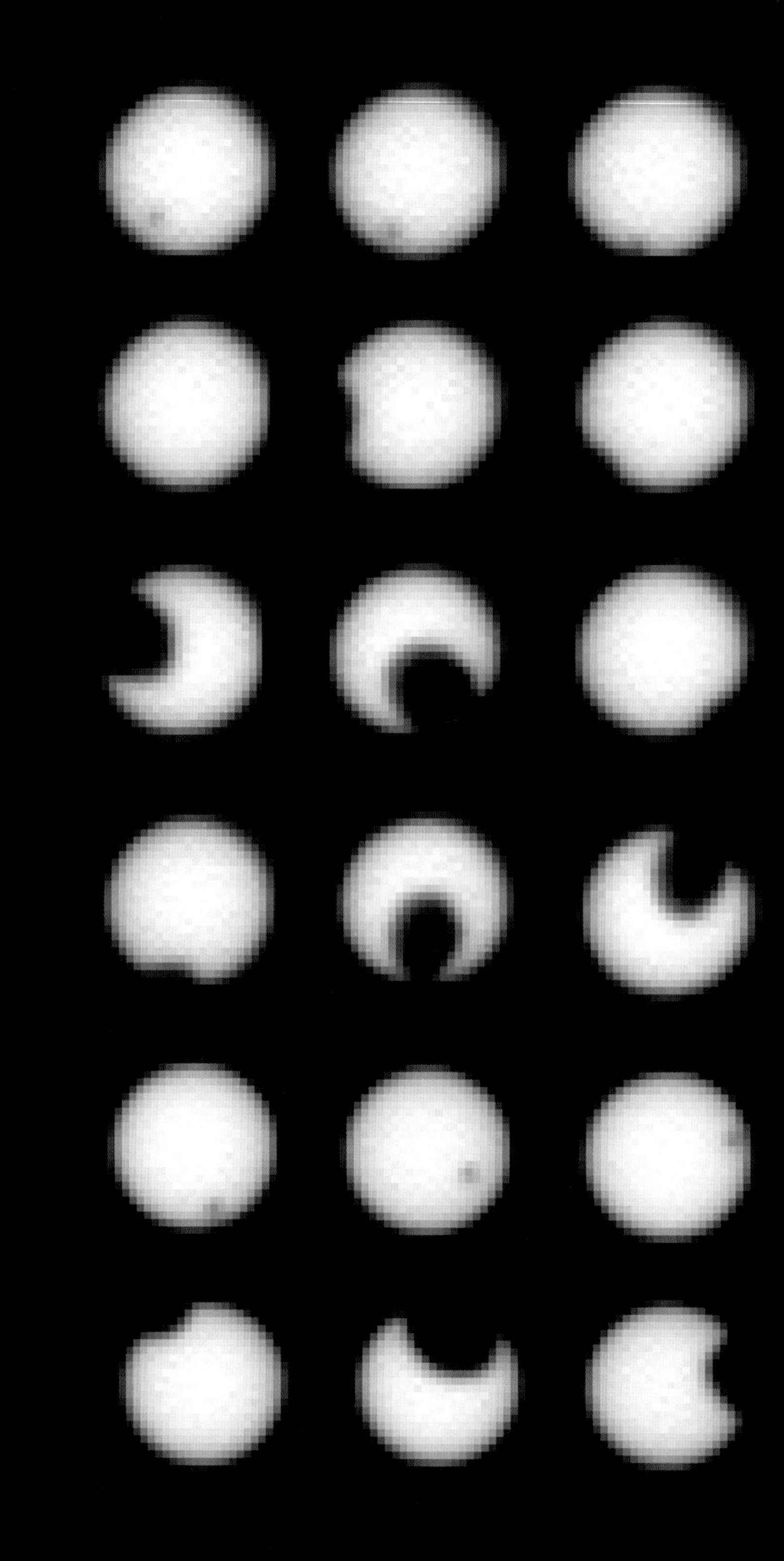

Time-series Pancam images of Phobos and Deimos passing in front of the disk of the Sun. Each row is a different "transit" event, with time running from left to right. (NASA/JPL/CORNELL)

process. Different parts of the landing site were photographed on different sols and at different times during each sol. Not only do the shadows change from one part of the panorama to another, but the sky color and brightness also change from hour to hour and from SOL to SOL because of variations in the atmospheric dust. The result is that when all of the images are merged into one big mosaic, the seams and brightness changes from image to image can make it look very different than if you were standing there, just looking around.

For more than a century panoramic photographers have known that seams or sharp color changes make it very difficult to give a perceived sense of visual reality to a landscape. For purely science analysis purposes, seams and other image artifacts might be acceptable or even desirable. But to give the scene a sense of realism, for many of our large rover color composite panoramas we removed the artifacts and attempted to simulate the view a person would see if all the images in the mosaic were taken on the same day, at the same moment. Simulating a view like this is based on a combination of science, past experience, and some artistic license. The science is there because the colors and contrast are never "faked." The average color of the sky or the scene is ultimately determined from images of the MarsDial acquired at the same times as the large panorama pieces. Experience comes in when large changes in color are observed across particularly bad image seams, and we have to choose which we prefer for the final product. Many of us involved in the mission have spent time viewing Mars through the eyepiece of telescopes large and small, and contributing to past efforts from *Viking* and *Pathfinder* to generate estimated true color views of the landscape from those data sets. So there is a sort of "shared perceptive consensus" on what the average color of different bright and dark parts of Mars should be to the human eye. Artistic license is employed when choosing whether or not to crop out parts of the rover in the foreground, or where to "cut" the 360° view so that it can be displayed in a flat, two-dimensional image format.

It's not an exact process, though, because the colors of the martian sky and landscape change significantly with time. Of course, people's perception of color can also vary dramatically from individual to individual. And color displayed on a television or computer monitor varies enormously, depending on how the monitor is set up. I've done the best I can to generate and share what I believe to be an accurate depiction of the colors of this alien world. Ultimately the true test of success will be the judgment of the first astronauts who go there and see the place for themselves.

Mission Success

Unlike the initial postcards, *Spirit*'s spectacular Mission Success panorama provided a much more complete view of the geologic setting around the rover's new home. The landing site was in a gently rolling plain, and about 5 percent of the surface was covered by rocks bright and dark, angular and smooth. Some rocks had bright wind-tails that helped identify the direction of the prevailing wind. Hills and mesas were common on the horizon. One set of hills to the east was only about 3 kilometers away. This would be only an hour or so walk for you and me, but it would represent an epic trek for a rover that we assumed would only travel 500 or 600 meters over the course of its maybe ninety-day lifetime.

Our cameras could faintly see brighter and darker bands on the ground as we looked out into the plains toward the hills. These are the dark streaks that we'd seen in orbital images, now viewed up close. To the north was a ridge that is the rim of a large crater. We could see from orbit that the crater has bright material on its floor, but we couldn't tell what it was from the rover. We could see several more of those enigmatic, bright-floored hollows; they are often ringed by rocks, but they have very few rocks inside. And we saw more of those dirty airbag parts at the edge of the lander.

What used to be pristine white fabric was now covered by fine-grained, reddish dust. It was a dirty, dusty, alien environment indeed.

Yet there is a sense of familiarity in these images from Mars. There's an "I've seen that place before" feel of looking out the window across a long drive in the desert somewhere. Rocks, hills, sky—it's all very Earthlike and comforting, in a way. But it's an illusion. It's 30 to 50 degrees below zero (°C or °F, it doesn't matter) on average out there; the air is almost entirely carbon dioxide, with only a trace of oxygen; and it hasn't rained in something like 2 to 3 billion years, if ever. There's not a hint of a cactus or a tortoise, or a wispy contrail from a passing jet. When you take a closer, more careful look at the landscape, you realize how truly ancient this terrain really is. The rocks have been carved and molded by sand and dust grains carried by the wind for billions of years. The ground is peppered with circular holes both large and small—impact scars formed when asteroids or comets crashed into the planet long ago. *Spirit* is driving on and over terrain that is as old or older than the *oldest* places preserved on the surface of the Earth. Some of the rocks we've examined are probably 3 or 4 billion years old, and *Spirit* driving past may have been the most interesting thing to happen in some of these places for the past billion years or more. The land is imbued with a sense of time, of age, of processes that have been at work for longer than even most geologists can conceive. By comparison, our home planet is young, geologically virile, and ever-changing. It is difficult and often risky for us to extend the geologic experience we've gained from living on such a young planetary surface to such an ancient place like Mars. It's only human nature, however, to want to feel like there's a little bit of home in every place we visit.

Six Wheels in the Dirt

Armed with a pair of superbly designed cameras and a strategy for choosing colors and compression levels that would allow us to get as much science as possible into the small daily mailbag of bits being sent back to Earth, the team embarked on what is arguably one of the most exciting and rewarding missions of planetary exploration ever conducted. The first seventeen sols of *Spirit*'s mission were focused on getting the rover safely off of the lander and onto the surface. "Six wheels in the dirt" was the mantra that we chanted. Many of us were stoked on the adrenaline rush of the successful landing, the ensuing international media spotlight, and a steady stream of caffeine, late-night ice cream, and new postcards sent back by our robotic friend on Mars. My camera team's job during those early days was to help acquire and process these initial "reconnaissance" postcards of the landing site, to help identify obstacles or opportunities related to getting off the lander, to help characterize the status of components on the rover deck and the lander, and to help identify potential rock and soil targets for detailed up-close studies once we got the rover on the ground.

In the initial pictures we could see several large obstacles around the lander. It looked like some of these might prevent the rover from driving off in an easy fashion. Were these rocks? Airbag materials? We didn't know what kinds of danger they posed to the rover. The mood was tense as the commands were sent to spin the rover around on the lander so that it could drive off a secondary "back door" ramp to avoid the obstacles (which turned out to be puffed-up airbag parts). This had been tested on the Earth, but really doing it up there was a lot more stressful. On *Spirit*'s twelfth day on Mars, we completed the final 35 cm of our landing, rolling off the lander and onto the surface. When the first images came down looking back on the lander from our new vantage point on the ground, there were applause and tears and hugs and glee all over again.

We fired up the Pancams for some color imaging of the lander and shot what has turned out to be one of the most stunning and emotional postcards of the entire mission. This so-called "Empty Nest" panorama shows our lander, now just a dusty, dirty, and empty cocoon, framed against the backdrop of the plains of Gusev and the distant hills on the eastern horizon. Many of the engineers were interested in seeing firsthand the detailed final configuration of the lander and the airbags. They knew that it had all worked, of course, but they also knew that future missions might benefit from a high-resolution after-the-fact study of a successful design. Many of the scientists were interested in the detailed color views of the soils and rocks close to the lander and the plains and hills in the distance. We were still trying to figure out what kind of place our new home really was.

Most of all, though, the emotion in this panorama came from the sense of pride and purpose in the people who built the equipment. Thousands of people from laboratories, universities, and companies all over the world were involved in building complex machines to do a risky, gutsy, seemingly impossible job: land rovers on Mars and do science with them when we got there. Here was this stunning image of some of their handiwork, now sitting out there forever in the martian sunshine, having done its job perfectly. From the biggest brackets, hinges, and pieces of aluminum to the smallest wires, cable ties, and airbag stitches, there are hundreds and hundreds of little engineering adventure stories locked up in this one image. From the smallest pebbles and dunes to the largest boulders and distant hills, there were hundreds and hundreds of little science adventure stories waiting to be discovered. I made a big print of the *Spirit* lander image and had as many people who were involved in the mission as I could find autograph it.

Columbia's Hills

The hills that we saw to the east were exciting but also a bit demoralizing. It was exciting to see interesting geologic features that told us that we must have landed in a geologically diverse place, but it was discouraging to think that we'd probably never be able to get to those hills. *Spirit* and *Opportunity* were designed to drive perhaps 500 or 600 meters over their estimated 90 SOL lifetimes on Mars. These hills, though, were more like 3,500 to 4,000 meters away. Hauntingly familiar hills had been seen in the distance at the *Mars Pathfinder* landing site back in 1997, but we'd only been able to drive that mission's *Sojourner* rover a few tens of meters away from its lander. It felt like *Spirit* was adrift on a vast ocean, in sight of land but without enough wind in the sails to steer the ship toward the most interesting port.

One thing that we could do about the hills in Gusev as a team was to name them. We had landed on Mars just under a year after the space shuttle *Columbia* had been destroyed on reentry in February 2003. The loss of seven courageous explorers was a human tragedy and an enormous setback for NASA's human exploration program. Many of us were deep into final rover testing and launch preparations at Cape Canaveral when *Columbia* and its crew perished. We felt a poignant sense of kinship and grief with our colleagues at neighboring Kennedy Space Center and all across NASA. As one tribute to the crew of *Columbia*, the rover team designed a special plaque for the back of each rover's high-gain antenna to commemorate the astronauts and their ultimate sacrifice in the name of exploration. As a second tribute, the landing site itself was named the Columbia Memorial Station. This followed a tradition established on previous Mars lander missions: The *Viking 1* landing site was named the Mutch Memorial Station, after the late planetary geologist and lead *Viking* imaging scientist Dr. Thomas A. ("Tim") Mutch. The *Viking 2* landing site was named the Soffen Memorial Station after the late *Viking* project scientist Gerald Soffen. And the *Mars Path-*

In the closest thing to a Pancam self-portrait, the shadow of *Spirit*'s mast stretches across the plains in this SOL 111 Navcam image. (NASA/JPL/CORNELL)

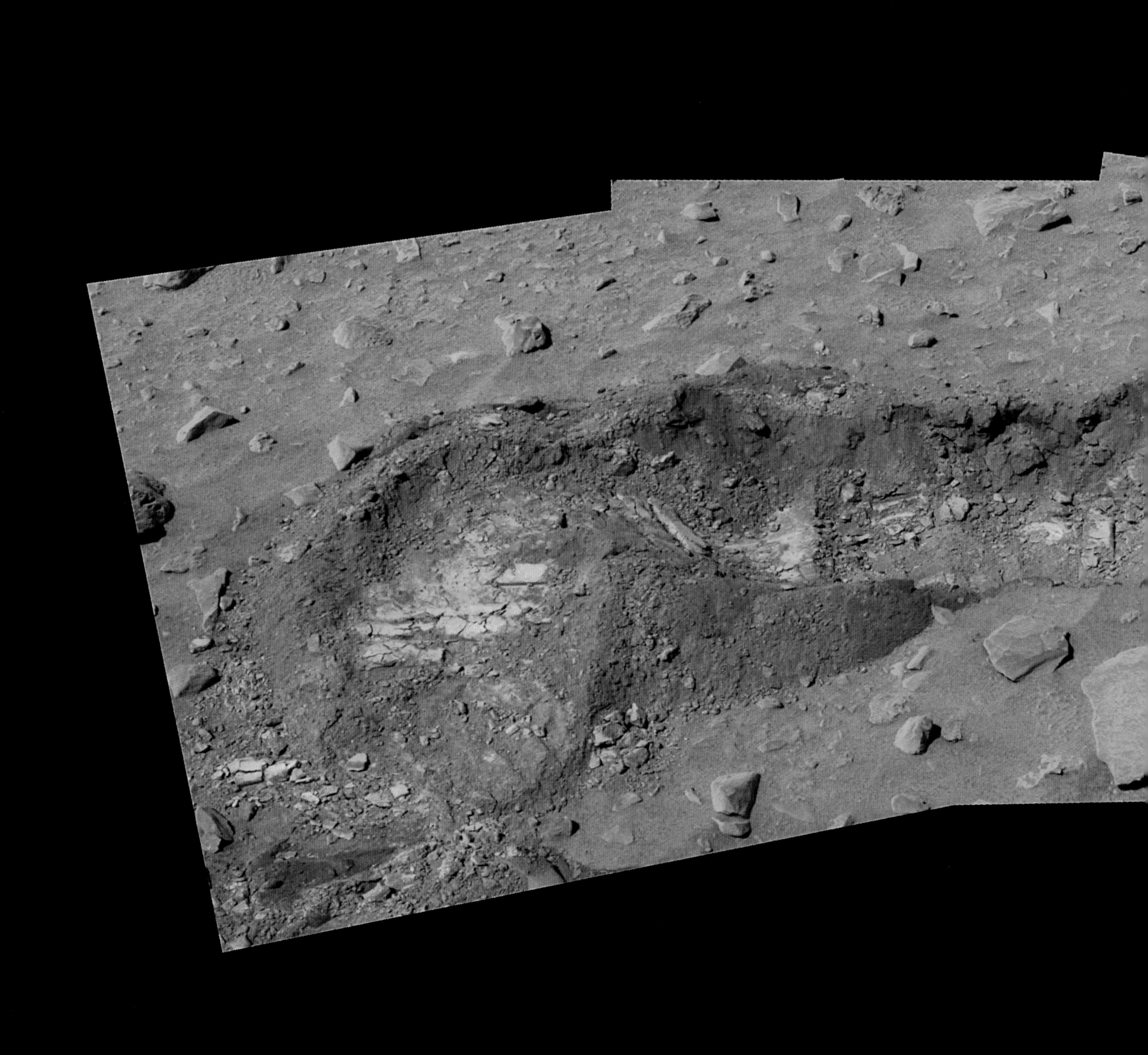

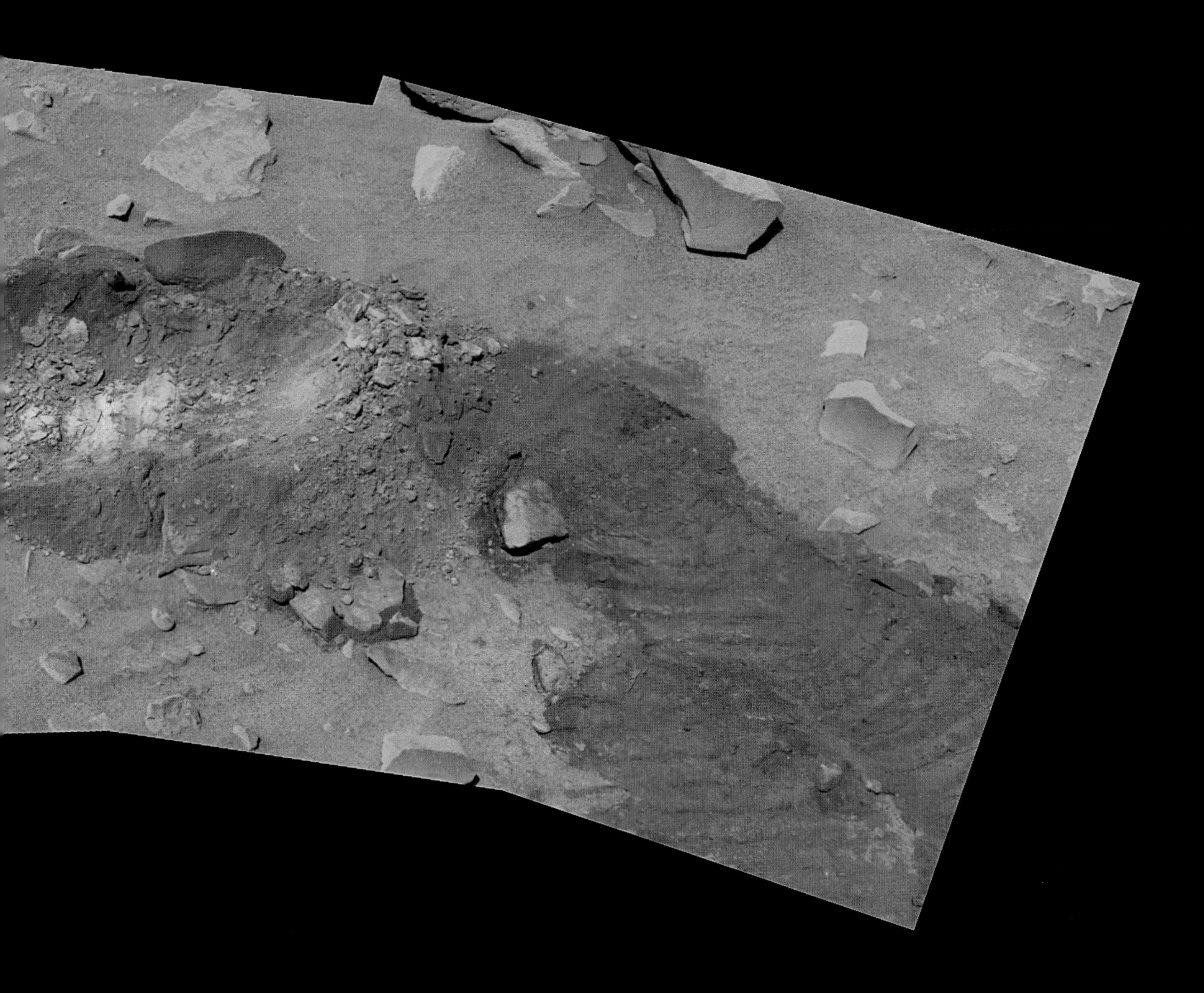

A 1.5-meter-wide trench dug by *Spirit*'s wheels in the Gusev plains, SOL 116. (NASA/JPL/CORNELL)

Spirit's last postcard of the Columbia Hills from the plains, SOL 149. (NASA/JPL/CORNELL)

finder landing site was named the Sagan Memorial Station, after the late Cornell astronomer and leading popularizer of science Carl Sagan. As a final tribute, the set of seven spectacular hills that we could see to the east of the *Spirit* landing site was named the Columbia Hills. Each peak was named for a member of the crew. The first six are named after Pilot William C. McCool, Payload Commander Michael P. Anderson, Mission Specialists David M. Brown, Kalpana Chawla, and Laurel Blair Salton Clark, and Payload Specialist Ilan Ramon. The seventh and tallest hill was fittingly named Husband Hill, after *Columbia*'s commander, Rick D. Husband. In a similar commemorative act, a second set of three hills visible in the distance were named the Apollo 1 Hills, with individual peaks named after astronauts Virgil I. ("Gus") Grissom, Edward H. White, and Roger B. Chaffee, all of whom perished in the *Apollo 1* fire in 1967.

No Beeps and No Flash

Once we got *Spirit* on the ground, our goal was to make some quick "contingency" measurements just in case the rover didn't survive more than a few days. This kind of paranoia is common in the space exploration business, when you never know which day will be the last one before your instrument or spacecraft breaks down. Even the *Apollo* astronauts followed such contingency plans, quickly collecting and stashing some rock and soil samples as soon as they stepped out onto the lunar surface, just in case they had to jump right back in the spacecraft and abort the mission for some reason. In our case, we acted like there was a sniper sitting in those hills off in the distance, taking shots at us as we trundled around doing our work. We had to get as much done as possible before we got hit. We made the soil contingency measurements and quickly identified a suitable nearby rock, which we majestically dubbed "Adirondack" even though it was only about 30 cm (a little over a foot) across. We drove the rover a few meters over toward Adirondack and deployed the rover's arm instruments onto it to make our first chemical measurements of a rock on Mars. It was then, just a few days after driving off the lander, that *Spirit* suffered a highly publicized and nearly fatal computer malfunction.

All we saw back on Earth was a series of strange, garbled radio signals. Even more confusingly, the short, familiar radio "beeps" that the rover radioed back to Earth occasionally to signal its health were not being seen when expected. Then, ominously, no beeps were seen at all. Silence.

The reasons why *Spirit* nearly died are complex, but the problem boiled down to an over-full memory and some sneaky little software bugs that weren't found during testing before launch. Each rover's brain is a computer that runs an operating system similar to a personal computer's but is optimized for the special demands and constraints of spacecraft in distant and alien environments. Each rover computer has a memory and data storage section based on the same kind of "flash" memory chip technology that is found in many small consumer electronics equipment, such as digital cameras. When we take pictures or make other measurements, the data are stored as files in a file system in the flash memory, just like files on a PC. Later, they are converted by the computer into the binary packets of ones and zeros that are sent back to Earth. We can store about 256 megabytes of data in each rover's flash memory. Some of the memory is devoted to other functions, though, so we can only use about half of that space to store scientific data. After the files are received safely on Earth, they are deleted from the flash memory to make room for new data. This makes "flash management"—monitoring the volume of new data compared to how much old data was deleted—an important part of the team's daily rover planning activities. If we've accidentally overfilled the flash, the operating system is programmed to automatically delete some of the older, possibly not yet downlinked, data.

One of the undiscovered software bugs involved the way files were being deleted in flash. In computer systems, files usually have two parts: the data in the file, and a "pointer" that tells the computer exactly where the file is stored in the computer's memory. It turned out that when a file was being deleted on *Spirit*, the data part was being deleted but the pointer wasn't. So, new files were being "pointed" to start deeper and deeper in the flash memory. The second bug, which compounded the first one and caused all the troubles, was that the operating system software started to lose track of the pointers, too. As the days went on and more and more files were written into flash, eventually the operating system became very confused and sent *Spirit*'s brain into uncontrolled spasms. *Spirit*'s preprogrammed self-preservation software kicked in and recognized that there was a problem. It commanded itself to reboot to try to fix the problem. However, the operating system was still confused. As soon as it rebooted, the problem appeared again and the rover shut itself down again. *Spirit* was caught in an infinite loop—rebooting, crashing, rebooting, crashing—for hours and hours and hours. All that time the rover was slowly draining its precious battery power. We didn't know any of that, though.

Diagnosing and repairing *Spirit*'s computer from across the solar system was not easy. Glenn Reeves and his team of JPL system and software experts who programmed the rovers were working day and night. They were in the dark all the time, however. They had no reliable communications and only a tiny amount of garbled data with which to work. They knew that the clock was ticking as the batteries ran down. The mission engineers had to get control of the rover quickly or *Spirit* would die.

In a strange way, though, it was one of the most thrilling and dramatic experiences in the entire mission. One of the hallmarks of NASA's history of space exploration, perhaps best exemplified and immortalized by the dramatic rescue of the *Apollo 13* astronauts, is the way that a small group of smart, clever, and motivated people under intense personal and public pressure can combine their brains and experience to solve what appear to be impossible problems. I witnessed this firsthand at JPL during the *Spirit* computer crisis, and it was an amazing and inspirational sight. Poring over piles

LEFT TO RIGHT
Martian rock fragments near *Spirit*'s front left wheel. (NASA/JPL/CORNELL)

Spirit SOL 163 Microscopic Imager photo of part of Pot of Gold, showing a close-up view of the strange textures of this rock. The image is 3 cm across. (NASA/JPL/USGS)

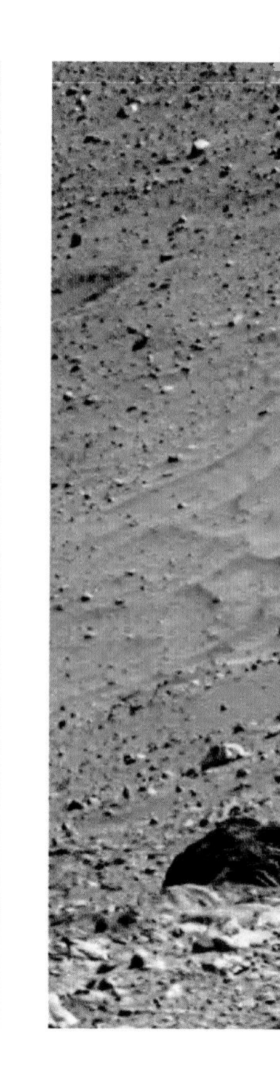

of computer code and tiny snippets of garbled data, the team figured out that *Spirit* was caught in a loop rebooting itself. If it had been a PC on your desk at work, you would have just pulled the plug and reloaded the operating system from your set of original install CD-ROMs. But this was impossible to do, of course, from 100 million miles away.

Instead, the team used a "back door" way to break into the loop. They had devised this years earlier, when the code was being written, just in case something like this ever happened. That someone had thought to program such a contingency for such a low probability software glitch is amazing to me. It's typical, though, of the kind of attention to minuscule details and backup plans that is often required in spacecraft missions. It worked, and the team was able to restore limited communications to *Spirit*. With new data came a more detailed diagnosis and eventually a treatment plan. The file system would be reformatted, thus allowing the computer to reboot successfully. The software bugs were fixed and new software was radioed to the rover. Within about ten days, *Spirit* went from critical condition in the intensive care unit back to completely healthy as the repaired software was tested and verified. We had lost some of the data that had been stored in flash, but it was nothing that couldn't be reacquired. With healthy instruments and successful interplanetary rover brain surgery complete, we were once again ready to get back to some photography.

Software folks aren't often recognized as heroes, but Glenn and the software and systems teams saved the mission with their detective work. Looking back, we probably could have detected the software bugs during our earlier testing. Unfortunately, the crash happened about seventeen days after landing, which was about three days longer than the longest prelaunch test that we had been able to run with the rovers and software in the same configuration that they would have once

they were on Mars ("Test as you fly, and fly as you test" can only work for so long...). Fortunately, the experience with *Spirit* helped us fix the same bugs on her sister, *Opportunity*, which was landing on the opposite side of Mars just about when *Spirit* came back to life.

"The Basaltic Prison"

Once *Spirit* was pronounced fit and ready again for active duty, we resumed our first upclose investigation of Adirondack. *Spirit*'s arm had been resting there for nearly two weeks. This was our first real chance to use the two spectrometers, the microscope, and the rock abrasion tool (RAT) on a rock in the way that we had intended for them to work together. First we measured the natural, dirty surface of the rock with the APXS and Mössbauer instruments and took pictures of it with the Pancam and the microscope. Then we brushed the dust off with the RAT and took the measurements again of the "clean" natural surface. And then, finally, we ground a hole 5 mm deep into the rock with the RAT and made the measurements again, this time sampling just the pristine rock without any of the "dirty" dust or coatings. The differences among the natural, brushed, and ground photographs and other measurements are interesting. Adirondack is actually a dark gray rock, but enough of the famous martian dust has settled onto it over time, so it appears red at first glance. In fact, *Spirit*'s wanderings around the lander during her first month on Mars revealed that most of the rocks are just as dusty. It was a good thing we had the RAT so that we could brush and grind our way under that dust and figure out what the rocks were really like. The RAT was the stylist on our extraterrestrial photo shoot.

LEFT TO RIGHT
False-color (for mineral analysis) Pancam photo of twin RAT holes ground into the soft rock named Wooly Patch. West Spur of the Columbia Hills, *Spirit* SOL 200. (NASA/JPL/CORNELL)

Pancam false-color photo of a group of Columbia Hills volcanic rocks called "Toltecs," *Spirit* SOL 220. (NASA/JPL/CORNELL)

Rocks and sand dunes along the flanks of Husband Hill. *Spirit* Pancam SOL 236 false-color view. (NASA/JPL/CORNELL)

Sparkly, probably sulfur-rich soils in the West Spur of the Columbia Hills. *Spirit* Pancam false-color photo, SOL 165. (NASA/JPL/CORNELL)

FOLLOWING SPREADS
88-89
Another view of the *Spirit* Santa Anita panorama, in the plains near the base of the Columbia Hills, SOLS 136-141. Image cut in two parts placed one above the other. (see also PAGES 68-69) (NASA/JPL/CORNELL)

90-91
Spirit's Bonneville Crater panorama, SOLS 68-69. (NASA/JPL/CORNELL)

92-93
Spirit's "Whale" panorama, showing ridges and outcrops studied along the flanks of Husband Hill, SOLS 497-500. (NASA/JPL/CORNELL)

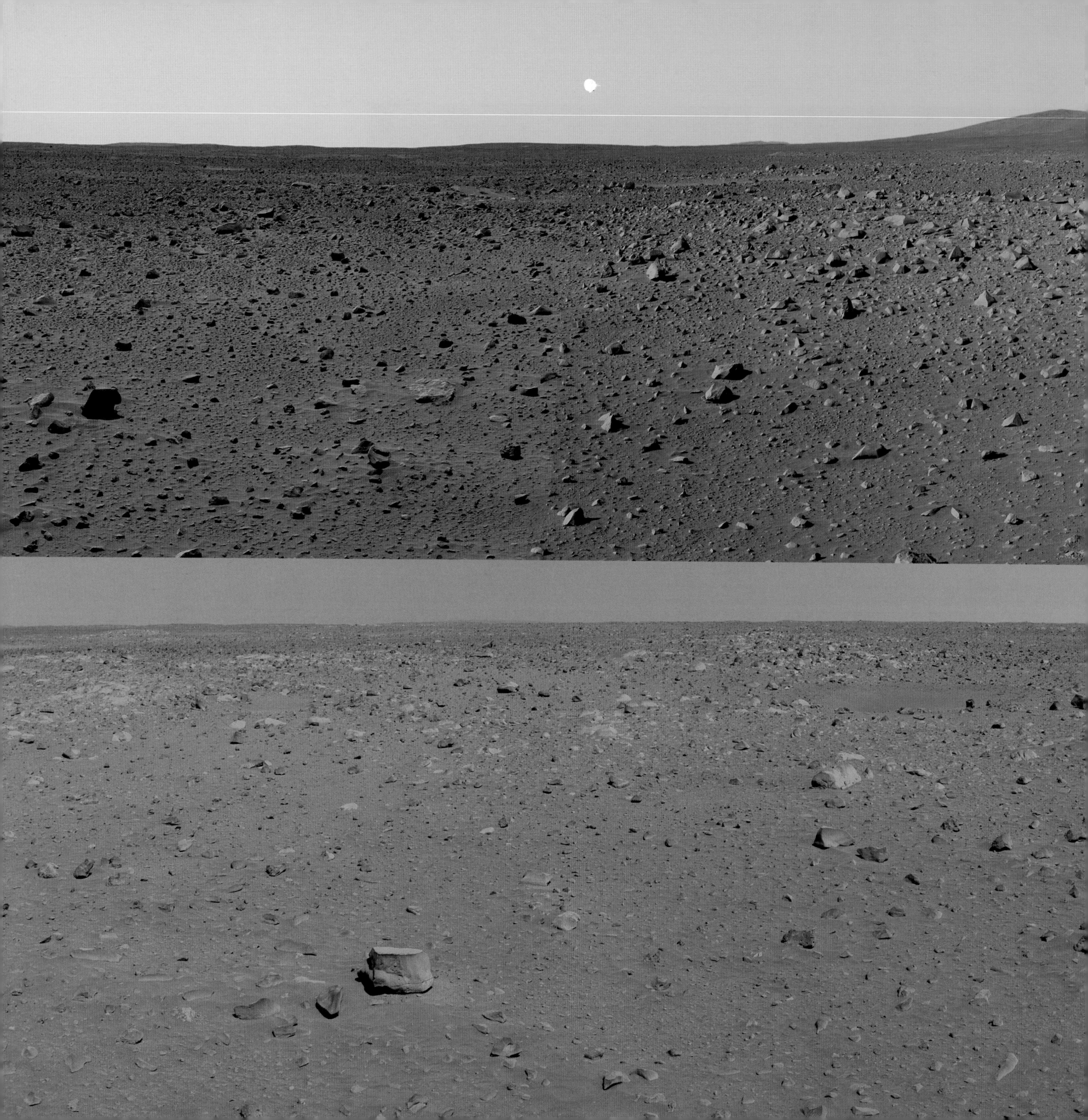

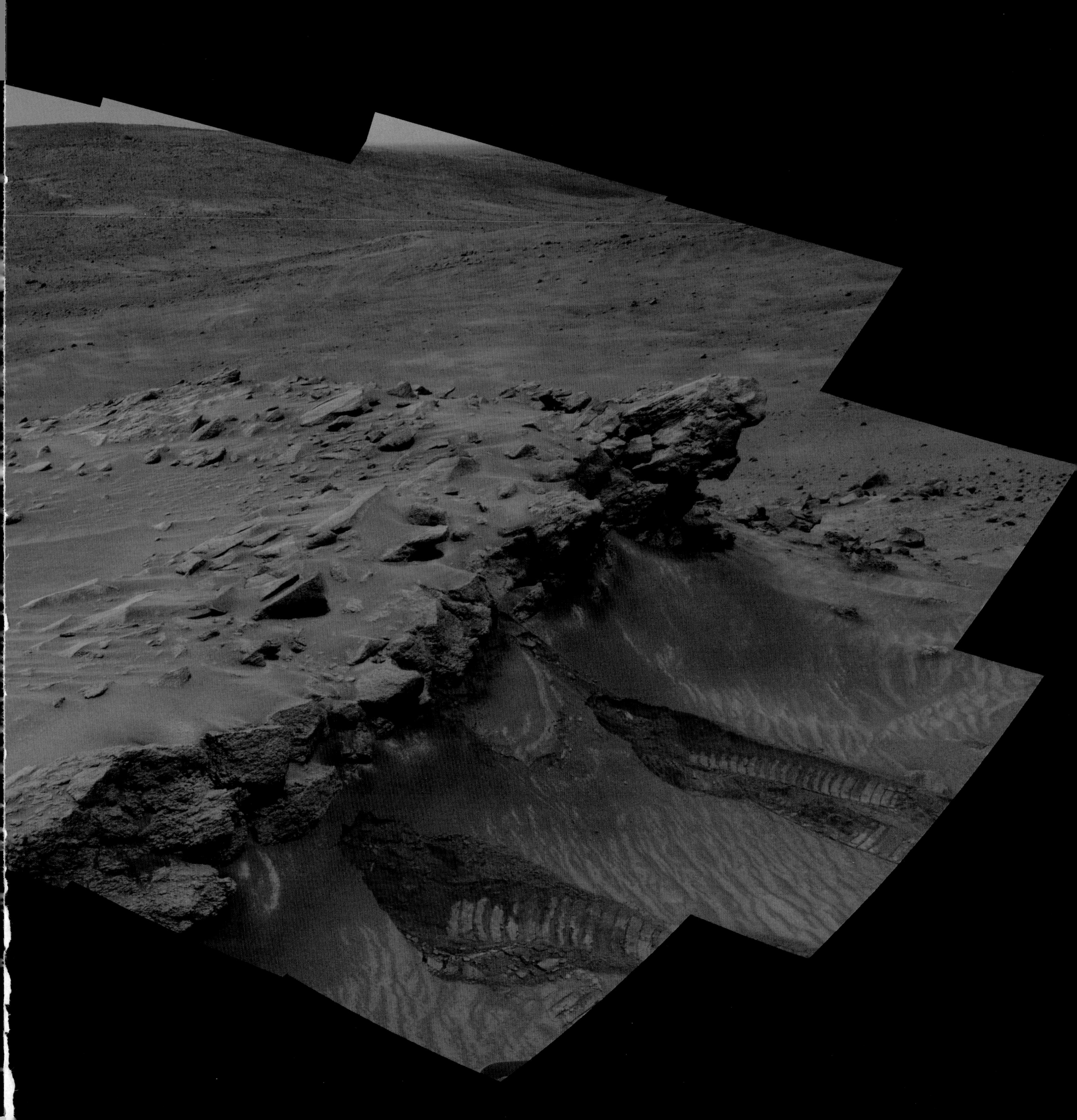

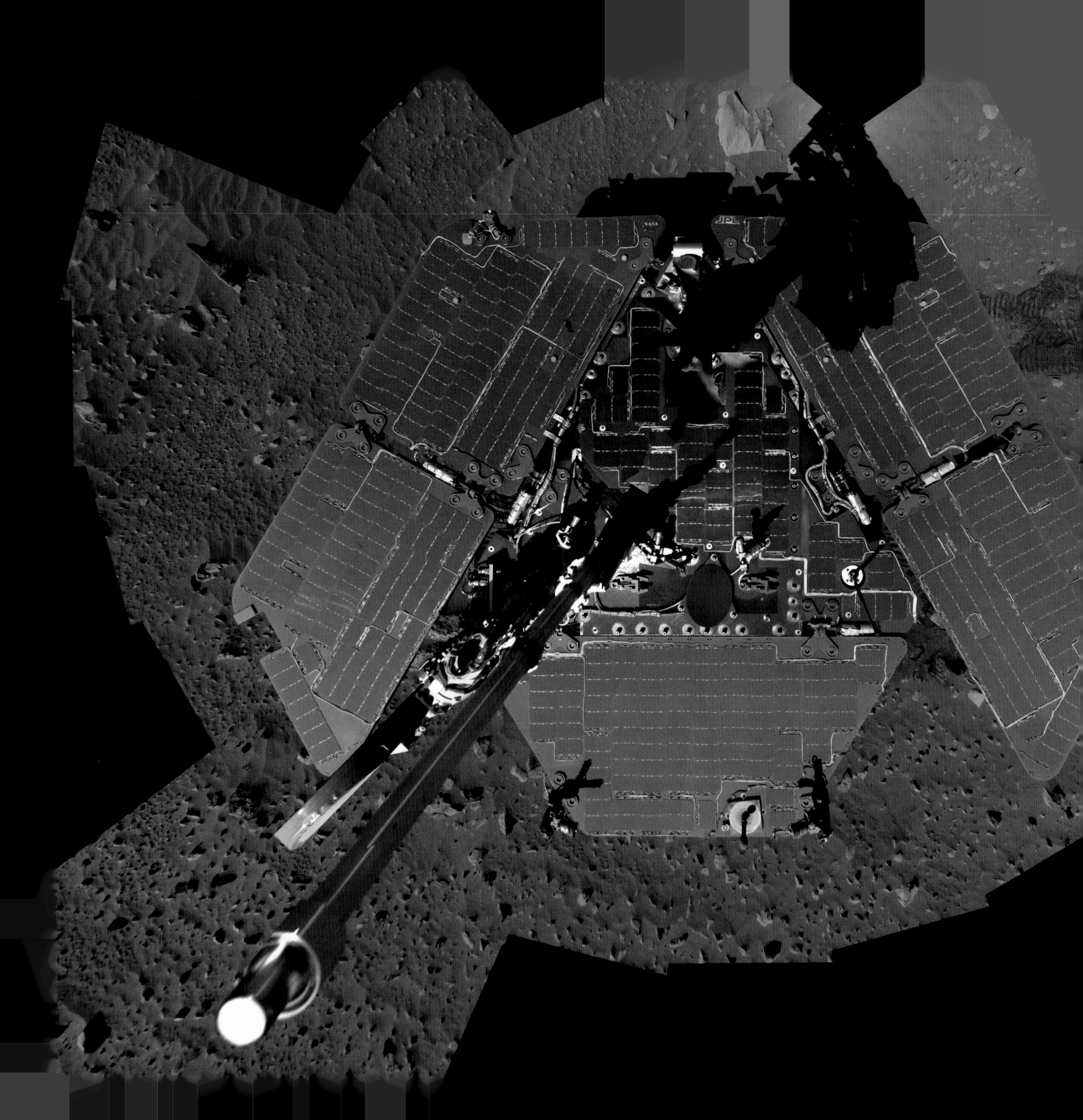
NASA
JPL

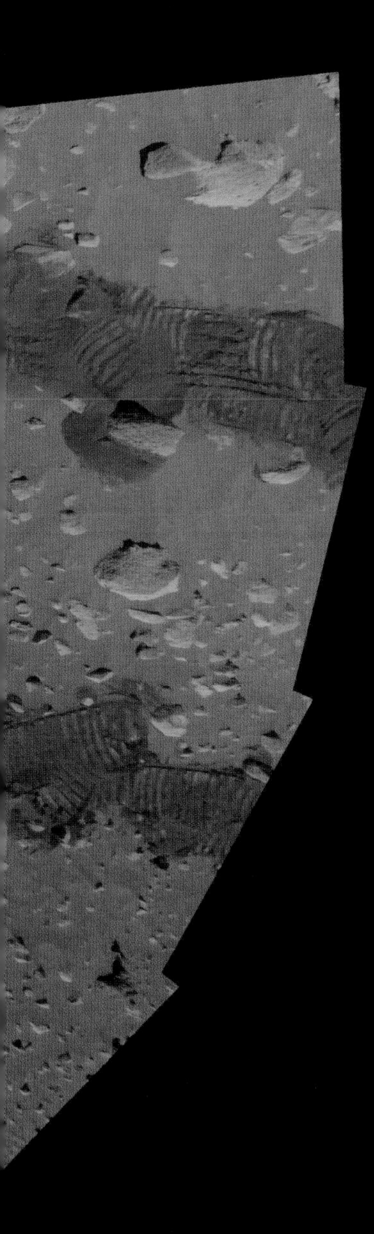

Spirit's SOL 329–330 "self-portrait" acquired by pointing the Pancam down toward the deck and taking photographs in a circular pattern around the base of the mast. Special image processing creates the illusion of the viewer hovering above the dusty rover's deck. (NASA/JPL/CORNELL)

PREVIOUS SPREAD
A triptych of *Spirit*'s 360° Husband Hill Summit panorama, including full coverage of the rover's deck. The panorama was acquired on SOLS 583–586 and consists of 653 separate Pancam images. (NASA/JPL/CORNELL)

THIS SPREAD
Sunset on the Gusev Crater rim. *Spirit* Pancam SOL 489 enhanced color photo. (NASA/JPL/CORNELL/TEXAS A&M)

FOLLOWING SPREAD
Microscopic Imager postcard of the heavily weathered Columbia Hills rock named "Keystone," *Spirit* SOL 439; field of view is about 20 x 12 cm. (NASA/JPL/USGS)

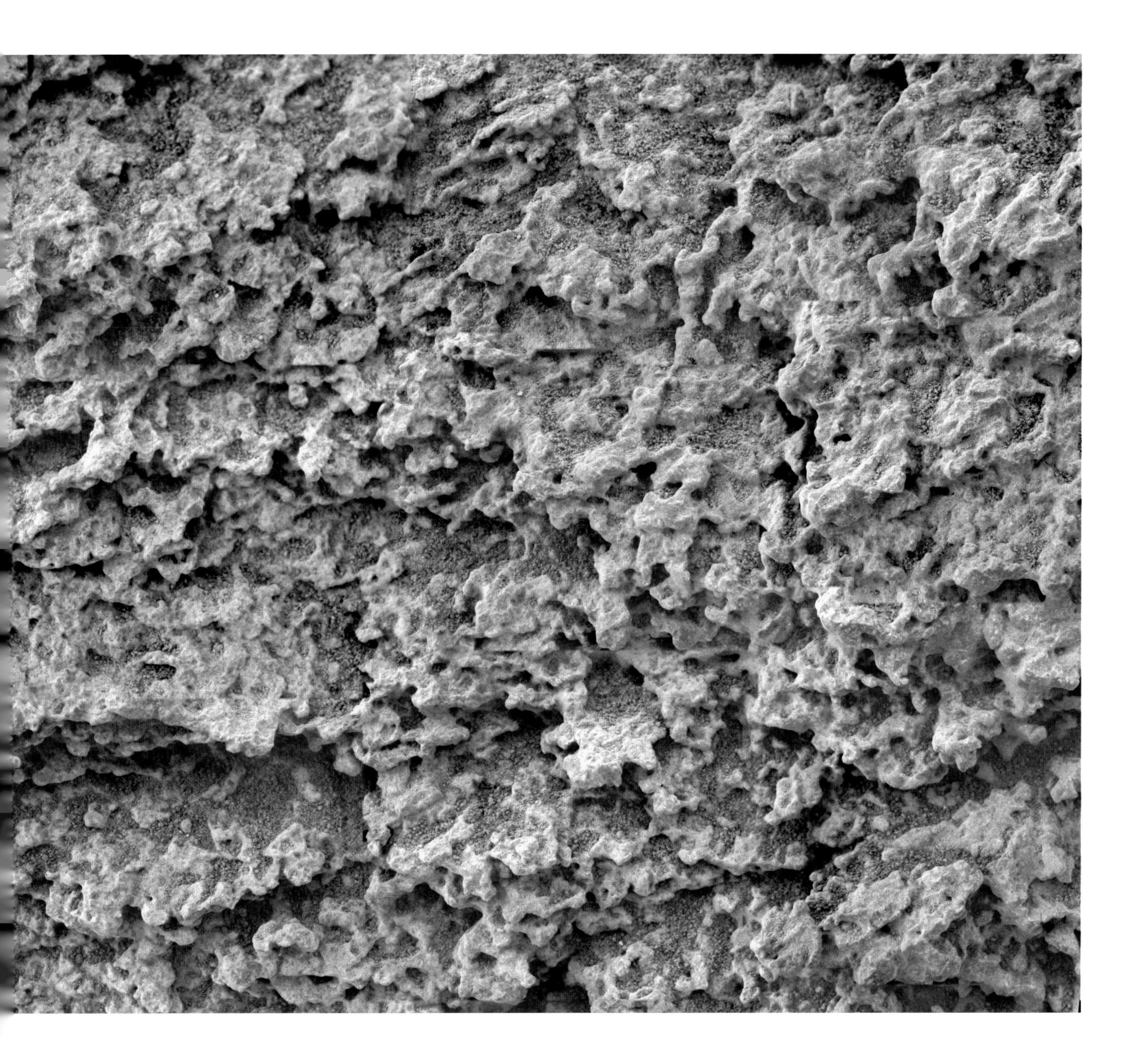

Close-up Pancam false-color (for mineral analysis) photo of rocky outcrop at Larry's Lookout, *Spirit* SOL 492. (NASA/JPL/CORNELL)

A trio of wind-carved rocks. *Spirit* Pancam false-color photo, SOL 506. (NASA/JPL/CORNELL)

What we found from our chemical and mineral measurements was that the rocks where *Spirit* landed in Gusev are all volcanic. They're similar, in fact, to a kind of common volcanic rock on Earth called basalt. It was exciting that the instruments were working so beautifully and that we were able to identify specific kinds of rocks and minerals with our robotic geology toolkit. However, the results we were getting were not what we had hoped to find. We went to Gusev because pictures from orbit showed that it might have once been a lake. As we roved around, though, we didn't see any evidence of sedimentary rocks or ancient shorelines or water-formed minerals. We didn't find any of the telltale signs that might indicate that the environment here had been different, possibly even more Earthlike. We found Earthlike rocks, all right. But aside from a few rare exceptions they were from old, bone-dry volcanoes and were what a geologist would call "pristine"—little evidence for water having been in, on, or even anywhere near them. Among the few exceptions were a few so-called "white rocks" that looked like they had thicker coatings of dust. Dust had apparently been "cemented" onto the rocks, possibly by the presence of a little water condensing on the dust over time. Some people got very excited about these white rocks, thinking that maybe they were the salts, carbonates, or other sedimentary rocks that we'd come looking for. It turned out, though, that they weren't really white at all. Once we took color pictures of them we saw that they were just brighter shades of red. When we ground into them with the RAT, we saw the same old dark, pristine volcanic minerals.

Even the dirt and sand that *Spirit* studied near the lander turned out to be just as bone dry and pristine. It appeared to be mostly made out of small chunks and fragments of the same kind of volcanic minerals that we were detecting in the rocks. In places where we could see the tracks from our wheels and cleats, it appeared that the weight of the rover was naturally digging down below a thin, bright, reddish upper-soil layer made from the fine dust that settles out of the air. A layer of darker, less red, pristine sand and rock fragments was often exposed by the rover wheels. This made it easy to follow our tracks, sometimes hundreds of meters back in the distance. *Spirit*'s wheel tracks were even seen from orbit, captured as a thin dark streak against the brighter plains materials in high-resolution images from the MGS Mars Orbiter Camera (MOC) by Mike Malin and his staff at Malin Space Science Systems. Over time, these tracks were observed to brighten and fade back into their surroundings, providing evidence for the active role that dust settling out of the atmosphere and blowing around plays in modifying the planet's surface even today.

The pristine soils and volcanic rocks were interesting, but they weren't what we had set out to find at this place on Mars. We hadn't come all this way just to be trapped for eternity in a sort of "basaltic prison," as some team members were beginning to call the place. We needed to try harder to see if there were places nearby that really did preserve evidence for those sediments or that would indicate that hypothesized warmer, wetter environment. Maybe we did land on an ancient lakebed, some thought, but the sediments and other deposits had been buried by eruptions from younger volcanoes. This was an idea that we could test by looking for areas where older rocks buried under the younger pristine volcanic basalts may have been dug up or exposed by natural processes. Indeed, our cameras showed *Spirit* was surrounded by such places—impact craters.

Off to Bonneville

We could see the rim of a 200-meter-wide impact crater off to the northeast almost as soon as we landed. We knew that such a big hole in the ground must have exposed quite a bit of subsurface material. Craters form from explosions when asteroids or comets smash into a planet's surface. They are like roadcuts for planetary geologists, providing a free glimpse below the surface but without having to do all the digging and excavating. Indeed, when we looked toward the large crater to our northeast—which we named Bonneville following a theme of naming features after dry lakebeds on Earth—we could see that it was surrounded by a raised rim and much more rugged, blocky terrain. It would be a long traverse traveling uphill for about 600 meters. It was as far as the rover was ever intended to go. But if we wanted to try to find different kinds of rocks, we had to get to Bonneville.

After about thirty sols' journeying we met Humphrey, a rather special rock. *Spirit* had roved and climbed, trundling over boulders and slipping on loose rocks and dirt. At Humphrey we dug a trench with the wheels, nuzzled up to the rock face, brushed and drilled below its dusty, dirty outer layers with the arm, and took detailed color and microscopic photographs of the rock's interior. We found tiny white cracks. It seemed we had found veins of weathered material formed as small amounts of water percolated through the rock. Finally! Some evidence for the interaction of rock and water at the landing site! Humphrey was the best evidence that we'd collected so far of water on Mars. The team was recharged and spurred on to continue the climb to the rim of Bonneville crater.

Cresting over the rim of Bonneville and seeing the interior of this huge crater for the first time was, in a strange way, both exhilarating and depressing. It was exhilarating to have worked hard as a team to complete a historic traverse across rugged terrain with a rover on another planet, and the view looking down into the crater was breathtaking. But what the stark view and our subsequent measurements *didn't* reveal was depressing. The crater was filled—nearly to the rim in places—with dust and sand. The rocks and soils that we analyzed were just more of the same dry, pristine volcanic basalts. No outcrops or layers were exposed inside the crater walls. No buried sedimentary rocks or different minerals were excavated and exposed by the impact. It was stunningly beautiful, but there was no scientific reward waiting over the rim. It was a bittersweet scene.

The Long Trek East

After making some reconnaissance measurements along the rim of Bonneville, we found ourselves at a crossroads. If we were tourists, our visa was about to expire. The rover and its cameras were not expected to last much longer. While we'd garnered an enormous amount of public and media interest in the mission, we hadn't achieved our scientific objectives. We knew where we had landed and took advantage of the fact that there were other missions and instruments, high above us in orbit, that had been taking measurements for us of our surroundings. The MOC, the cameras on the *Mars Odyssey*, and even the newly arrived European *Mars Express* orbiter cameras were obtaining new infrared and stereo images of *Spirit's* surroundings in an attempt to help us find more interesting terrains to drive to and explore. What these measurements, as well as our own images and other data from *Spirit*, showed was that we needed to head for the hills.

We saw the Columbia Hills right after we landed, of course, but soon learned that they were an impossible 3 to 4 km distant. This was six to eight times as far as we had hoped the rover could cover over its lifetime. The data coming in from the orbiters and our own Pancam views of the hills were tantalizing. The topography of the hills was rugged, and there appeared to be a sharp geologic boundary between the hills and the plains that we landed on. Maybe that meant that the volcanic rocks stopped at the hills, which rose up like an island of different material sticking up above a sea of basalt. Maybe. We could see subtle color variations in the hills that suggested layering or maybe even outcrops, but it was at the limit of resolution of the cameras. We couldn't

Pancam view of the grinding bits in the *Spirit* Rock Abrasion Tool or RAT, SOL 240. (NASA/JPL/CORNELL)

Spirit's robotic arm instruments, photographed on SOL 287. The RAT cable covers (with the American flag) on both *Spirit* and *Opportunity* are tributes to the victims and heroes of 9/11 and were fabricated from aluminum recovered from the World Trade Center wreckage in New York City. (NASA/JPL/CORNELL/HONEYBEE ROBOTICS)

Navcam SOL 287 image of *Spirit's* robotic arm, preparing to deploy onto rocks in the Columbia Hills. (NASA/JPL/CORNELL)

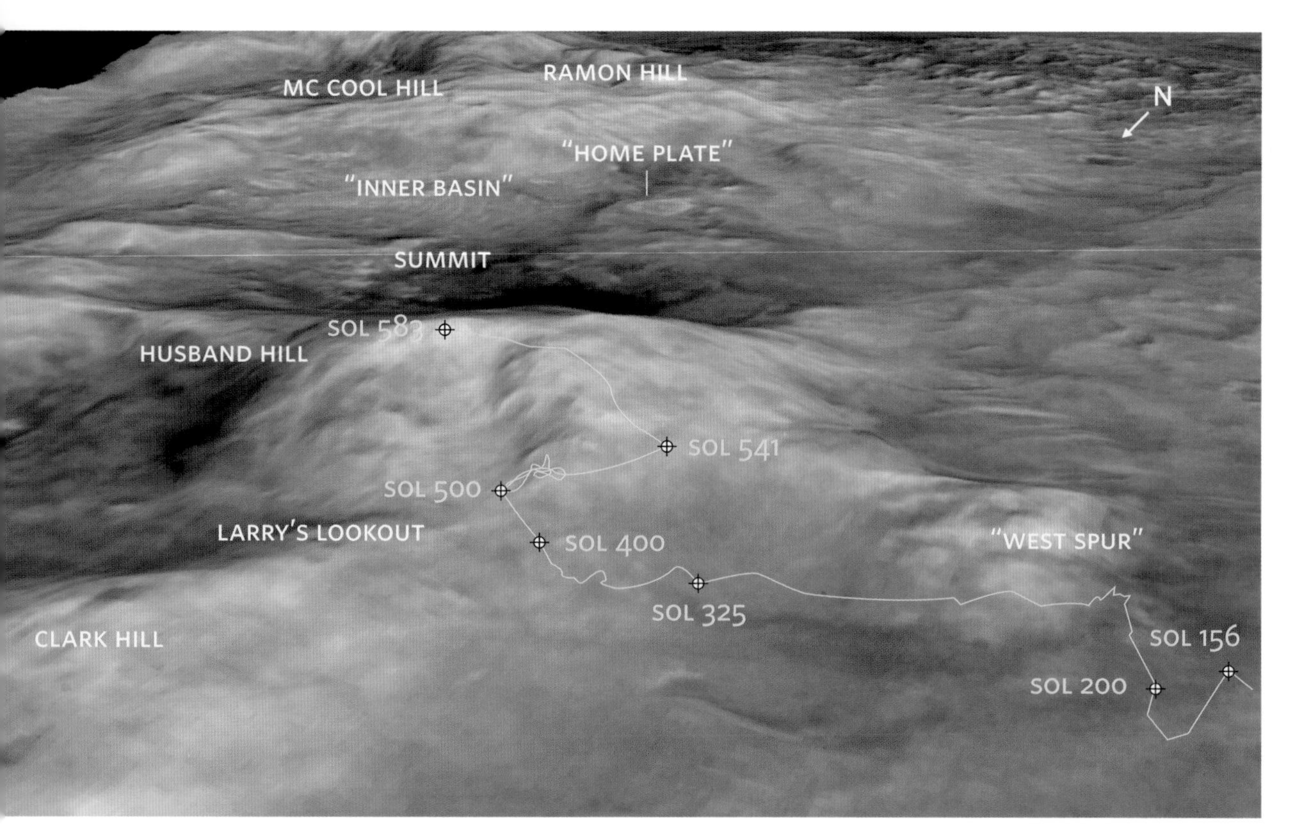

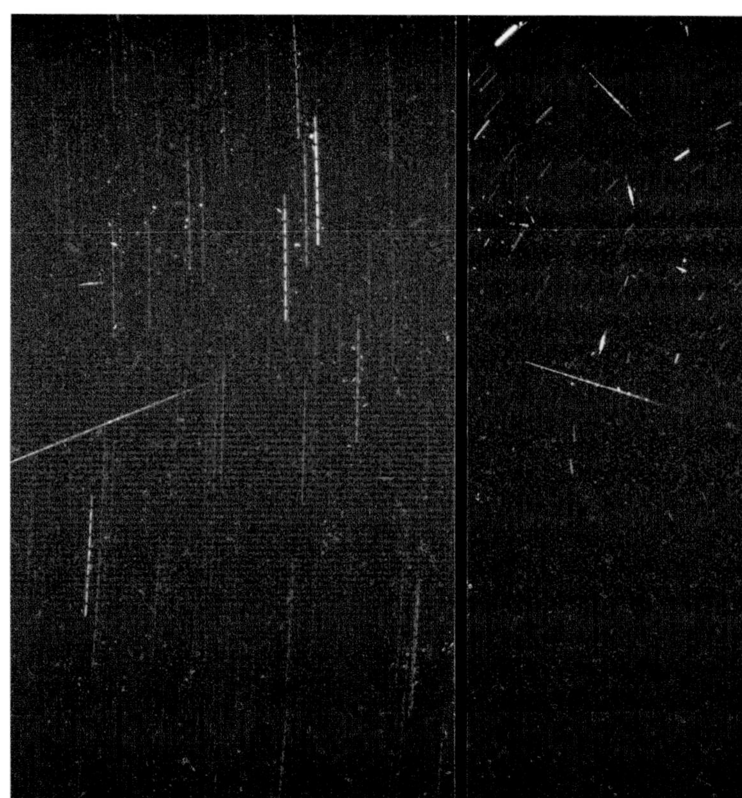

be sure. But what choice did we have? All around us was apparently the same kind of material. Yet just beyond reach was something else that was potentially very different. It would be a long and difficult trek, but *Spirit* and the instruments were in excellent health and the engineering team felt that the long drive was feasible. So we went.

With little fanfare but much trepidation, we set off for the hills on SOL 71. The driving was challenging for the engineering and mobility teams. We had decent evidence that the geology was probably not going to change significantly until we got to the hills, and we knew that somewhere that sniper was still out there taking shots at us. The decision was made to just drive, drive, drive as much as possible.

This turned out to be easy for some, but agonizing for others. On the one hand it was nice getting into a routine of sorts. Some of the team members came up with a four-sol series of repeating "drive quartets" that allowed us to occasionally do a little science while still keeping the odometer spinning. On the other hand, we were flying past (albeit at 4 cm/sec) so many countless rocks and soil deposits that we barely had time to even take pictures of, let alone analyze. I have heard some colleagues outside the team call this reckless, or even irresponsible. What if we drove past a dinosaur bone or a rock with fossils in it, as we had done during field tests with a prototype rover back on Earth a few years earlier? More realistically, though, it was a valid criticism to worry that maybe we weren't properly characterizing the variations in the geology and geochemistry of the plains as we bolted across them, so that when we arrived in the hills we could have lost the context for how they truly vary from the plains, if indeed they did. We worried about these issues, and tried to do the best we could to sample some of the chemistry, mineralogy, and geology of the plains while still trying to make good time on the road east.

We did stop for a few extended science campaigns during the long, lonely three months that it took to drive that roughly 2.6-km (1.6-mile) distance. At one stop we dug another trench with the wheels, 10 to 15 cm down into fine-grained dusty soils to see if we could detect any evidence for subsurface layering or compositional variations. Mostly we just found the same kinds of dusty, uppermost surface and darker, pristine subsurface that we had seen elsewhere. At another stop, we did detailed arm work on another so-called "white" rock, this time a large, flat, bright rock named Mazatzal. This turned out to be one of the most extraordinary rocks that we'd encountered yet, showing evidence for being coated by a thin and still poorly understood "rind" of weathered minerals possibly formed by the action of significantly more water than we saw in Humphrey. The water story was getting more compelling the closer we got to the hills.

LEFT TO RIGHT
Spirit's traverse from the plains of Gusev up into the Columbia Hills, depicted on orbital imaging and topography from the *Mars Global Surveyor* mission. *Spirit* reached the summit of Husband Hill on SOL 583 (August 24, 2005). (NASA/JPL/USGS)

Star trails, cosmic rays, and possible meteor streaks in the night skies above Gusev Crater. *Spirit* Pancam composite photos from three-minute exposures taken on the nights of SOLS 643 (left) and 668 (right). The SOL 668 image shows stars rotating around the martian south celestial pole. (NASA/JPL/CORNELL/TEXAS A&M/SSI)

Rocks, sand, and shadows in the late-afternoon Sun at the summit of Husband Hill, *Spirit* SOL 606 photo. (NASA/JPL/CORNELL)

Pot of Gold

Spirit made it to what became known as the West Spur of the Columbia Hills in July 2004, around SOL 157. During our drive across the plains, we took color images of the hills at higher and higher resolution to help us target the safest and most interesting parts to eventually drive up into and explore. Images of the West Spur, a smaller hill connected to the rest of the Columbia Hills complex, were used to choose it as our mountaineering entrance point partly because it appeared to be relatively easy to climb and partly because it is one of the closest places in the hills to *Spirit's* landing site.

The West Spur turned out to be a fascinating choice, because almost literally as soon as we started climbing we came across a bizarre and different kind of rock. We quickly found that indeed the geology and mineralogy were different from the plains upon which we had driven for nearly six months. In fact, the first interesting rock we found was so different that we named it Pot of Gold because we thought we must have found the end of the rainbow. Pot of Gold is a small rock, only about 10 cm across, sitting inside a much larger, round hollow of bright dusty soils near the base of the West Spur. We knew it was a strange and unique rock as soon as we saw it. Close-up examination with the cameras revealed what looked like a strangely eroded and etched surface, with "stalks" of rock jutting out and weird little nodules on their ends. It was unlike any rock we'd seen in Gusev. Some of the patterns looked familiar to the geologists on the team who had experience working with heavily eroded rocks on Earth. The clincher that we really *had* found a pot of gold was when the Mössbauer spectrometer detected the presence of coarse-grained hematite (Fe_2O_3) in the rock. Hematite is the iron oxide mineral discovered from orbital data that lead to the choice of Meridiani Planum as the landing site for Spirit's sister, *Opportunity*. Here in the Columbia Hills, as at Meridiani, it provided a beacon of sorts, indicating that liquid water may have been involved in the alteration and weathering of these rocks. This was a truly special moment, because we had found evidence, in a single rock, for both physical and chemical alteration that had a good probability of involving water. Pot of Gold vindicated our months' long trek across the plains to get to the hills. The fact that we'd found something so interesting almost as soon as we arrived meant that there were likely to be many more possibly water-related surprises in them thar hills...

As we continued to climb the West Spur we were thrilled to discover outcrops of rock. An outcrop is a holy grail to geologists interested in studying the history of a region. Some of the outcrops found in the West Spur even showed evidence for fine layering, which is another telltale marker of different and changing geologic environments. We found that the rocks we encountered in the hills were not pristine volcanic basalts. Instead, like Pot of Gold, they are weathered and altered rocks, with forms that appear to have been modified by the action of wind and—at last—containing minerals formed by the action of water and heat acting to modify all those original volcanic minerals. The Columbia Hills really did appear to be remnants of a more distant period in the martian past, when near-surface water may have been more abundant and when the climate may have been different. It took more than six months to find, but *Spirit* was finally among the kinds of materials that we had hoped to find when we selected the Gusev landing site years before. Getting to the hills was like starting a brand-new mission.

We took a path up the flanks of the tallest peak, Husband Hill. Often the rover was scrambling up the slopes, dislodging rocks and digging deep wheel tracks (mini-trenches) as it worked hard to climb uphill. The hard work had been causing problems with high motor currents on *Spirit*'s right front wheel. Special tests and driving strategies were developed by the engineers to try to minimize the impact of the wheel problems on the science that we could do in the hills.

Power Hungry

Around SOL 250 (September 2004), at nearly three times longer than *Spirit's* expected lifetime, decreasing power forced us to enter a new phase of the mission. Mars has seasons just like the Earth, although they're roughly twice as long as our seasons. Spirit had landed back in January 2004 in the middle of southern Mars summer. The Sun was high in the sky and nearly overhead for much of the day, and solar power was plentiful from the shiny, clean solar panels. Our traverse to the hills took the rest of the summer and much of the fall. Gusev crater is 15° south of the equator, and as fall became winter there the temperatures got much colder (down to a chilly -105°C to -110°C at night). The Sun was much lower in the sky and moving farther north every day. Less direct sunlight, combined with the fact that the solar panels were slowly getting dustier, meant that the rover's solar power was decreasing to alarmingly low levels. Much more of every sol's power budget had to go into just keeping the rover alive during the colder nights. Our daily science and driving operations became severely limited. We could drive or take pictures for only an hour or two a day instead of four or five, like we could earlier. When we drove or parked, we had to try to bask, lizard-like, with the panels tilted as much into the Sun as possible to maximize our power supply. Larry Soderblom from the USGS and other colleagues from JPL and the science team even came up with a clever way for us to figure out how to drive from place to place while maximizing our Sun basking. They drew up maps of so-called "lily pads" in the parts of the terrain where the slopes would give us the best sunlight. Those were the places that the rover drivers had to get us to every sol.

Before the rovers landed on Mars, most of us thought that what would ultimately limit their lifetimes was a combination of dust slowly settling out of the atmosphere onto the solar panels and the slow sinking of the Sun as summer transitioned to fall and then to winter. These factors combined would mean a slow, inexorable decrease in our available electrical power each sol. Eventually we wouldn't even have enough power to run the heaters at night. Some electrical component—maybe the radio transmitter or something similarly critical—would break in the extreme cold. A slow, cold, lonely death, but a life well-lived.

During the first few months of the mission we could see that the power trends were heading down in a way close to what we had predicted. In fact, the prediction was based on how the solar power had decreased during the *Mars Pathfinder* mission in 1997, so this wasn't too much of a surprise. It reinforced our desire to get to the hills soon and do new science there because we knew that the clock was ticking as power was going down. Once we were in the hills, though, we noticed that the power stopped dropping so quickly. It appeared that the atmosphere was generally less dusty during this particular Mars year than it had been while *Pathfinder* was operating. The amount of dust settling out of the atmosphere and sticking to *Spirit*'s solar panels was less than had been predicted. This was a stroke of good luck.

Dust storms form and grow all the time on Mars. Small local and regional storms are seen at almost all times of the year from the *Mars Global Surveyor* orbiter cameras. Sometimes during summer in the southern hemisphere, when the planet is closest to the Sun, they can grow into enormous clouds of dust that completely cover the planet. The last time this happened was in 2001. MGS was able to acquire exciting new scientific data on this uniquely martian weather phenomenon. The last global dust storm time before that was in 1971. In that storm, the thick red clouds of dust that covered the planet initially prevented the first Mars orbiter, *Mariner 9*, from mapping the geology of the planet. If a storm like that occurred during the rover mission, the dust blown around by the storm itself would get the camera lenses dirty but leave the rest of the rover operational. The atmosphere is so thin that the force exerted by the wind and dust blowing around would be tiny. The real problem would be that such a storm could blot out so much sunlight that we'd quickly find ourselves without enough solar power to survive.

With no big dust storms occurring near Gusev during the summer and fall, the seasonal effect of the Sun getting lower in the sky was the main source of our power drop. We hit the low point around the time of the winter solstice (September 2004), with our total available power down to only about one-third of what it had been when we landed. Mars has a very elliptical orbit, and when it's farthest

LEFT TO RIGHT
Close-up Pancam image of a rover wheel scuff mark, showing that the dark sandy Gusev soils are covered by a thin layer of bright reddish dust. *Spirit* SOL 73. (NASA/JPL/CORNELL)

Additional colorful sulfur-rich soil deposits were uncovered by the rover wheels in the Southern Basin on the drive toward Home Plate. *Spirit* Pancam false-color image; SOL 721. (NASA/JPL/CORNELL)

from the Sun it receives nearly 40 percent less sunlight than when it's closest. These were cold, short days for *Spirit*. Most of our effort was spent just staying alive, eking out a few hours of science observations or a short drive whenever possible. We even had to stand down from operations entirely for a couple of weeks during the time known as "conjunction." Mars was on the opposite side of the solar system from the Earth and was actually too close to the Sun (in the sky as viewed from Earth) to allow effective radio communications. It turned out to be a good time to rest, though, because both the rover and the team were going to need a breather before continuing on the long road ahead.

Once we passed through the depths of winter and the Sun began to climb back higher in the sky every sol, we started to see evidence from our own observations and those of the orbiters above us that the atmosphere was getting more dusty. However, we weren't seeing our power drop further. Rather, for reasons we didn't understand, we had appeared to reach some kind of equilibrium point in the dustiness of the solar panels. As winter turned into spring, the increasing sunlight was driving our power levels higher, back to the levels where we could operate longer and even do more complex science activities and drive sequences again.

Lucky Devils

Images from orbit showed that small dust storms and even tiny, mini-tornado-like dust devils were becoming more frequent. We were climbing higher and higher into the hills, heading for the summit of Husband Hill about 100 meters (300 ft) above the level of the plains. The climb provided some spectacular opportunities for landscape photography both in the hills and back toward the landing site on the plains to the west. We knew that those plains had dark wind streaks in them that were thought to have formed by dust devils sweeping the surface, and we actually saw a few small dust devils by serendipity in some images of the plains. We thought it would be interesting to try to take some special images designed to catch some dust devils in action. So we pointed the cameras down into the plains and acquired some time-lapse movies. The closest thing to a "movie" that our cameras are capable of is to take one frame about every five to ten seconds. We had to hope that we'd get lucky. And we did! Within days we started catching dust devils moving across the plains in the images. Over the course of months of occasional monitoring we saw hundreds of them. Most of these little cyclones were a few tens of meters wide and moving lazily across the plains at a few meters per second. Occasionally, though, we would spot a really big one—a hundred meters wide or more. As the plains got warmer with the advance of the seasons, it became clear that these mini-storms are a major way that dust gets moved around on Mars. This was a real coup for the atmospheric scientists on the team. Only a few of these dust devils had actually been seen back in 1997 at the *Pathfinder* landing site. Now we were able to make many detailed measurements of some of the most important weather phenomena on Mars today.

Dust devils were more than just a scientific curiosity to us, though. Indeed, they are probably responsible for helping *Spirit* live so much longer than its expected lifetime. As we watched the solar power change with time from both dust settling on the panels and the changing sunlight with the seasons, we would occasionally observe some bizarre events where our power levels would jump up by 10 percent or more almost suddenly during the daytime or overnight. We didn't have any good explanations for these (happy) jumps in power until we started seeing all the dust devils. One day, when we were parked up on a high ridge in the hills, we could even see that much of the rover's deck had been "swept off" overnight and was now much cleaner, presumably by the wind. Apparently, strong gusts of wind, possibly even from dust devils themselves, occasionally were blowing over the rover and cleaning the dust off of the solar panels for us. Some people joked that maybe some martians were actually cleaning the deck for us. Try as we might to take pictures of them with our cameras (even with a green filter), we never saw any such friendly, helpful aliens.

Scrambling in the Sulfur

As spring turned into summer, the Sun rose even higher in the sky. The solar panels stayed pretty clean, and our power levels soared. More than 500 sols after landing—getting close to a full Mars year later—our power levels were close to where they were when we landed. It was a power bonanza, and it enabled us to climb to the top of the hills. The climbing was often more like scrambling as the rover struggled up 20° to 30° slopes, dislodging rocks and slipping in loose soil patches. At one patch of particularly slippery soil, called Paso Robles, the digging action of the wheels accidentally unearthed (unmarsed?) some exotic whitish, sulfur-rich soil deposits that were unlike anything that we'd seen before. We think that these are deposits of iron and magnesium-rich sulfate minerals that are just below the surface there, but we're not sure how these minerals got there or why. On Earth, these kinds of deposits often occur in places where groundwater is interacting with sulfur-rich volcanic gases or fluids, but there aren't any active volcanoes (that we know of) on Mars today, so maybe these are the remnants of those kinds of water-related processes having acted at this part of Mars long ago. It was a puzzling discovery.

We had actually driven right past the Paso Robles soils without even noticing them. Typically, of course, we were focused on the road ahead and tried to come up with both safe and scientifically interesting routes toward the next waypoint in the long-term plan, which was to reach the summit of Husband Hill. Along the way we maintain the flexibility to stop and investigate new or compelling rocks or soils. These were usually things that we saw ahead of us from a previous drive location. In the case of the Paso Robles soils, we never noticed them as something interesting *ahead* of us, because they weren't exposed for view until the wheels had dug them up and they were *behind* us. Even the rear hazard cameras weren't much help, because their black-and-white pictures didn't reveal the distinct yellowish-white color of these interesting deposits. Luckily, there were enough people on the team who are interested in the properties of both pristine and disturbed soils that we try to occasionally take color Pancam pictures and Mini-TES infrared spectra of our wheel tracks as part of routine traverse monitoring activities. These kinds of images weren't considered to be particularly "tactical" data products needed for each day's driving decisions, and typically they were stored in a lower-priority part of the rover's flash memory. These lower-priority images can take many days to rise to the top of the queue for transmitting back to Earth. The result was that in the case of the color images of our tracks in Paso Robles, we didn't really notice these remarkable sulfur-rich soils until several days after we had driven past the area.

As soon as we saw what we'd uncovered, though, we did the rover equivalent of slamming on the brakes and pulling a U-turn. We headed back to Paso Robles to conduct the detailed chemical and mineral work and other measurements that could help us understand what was going on. It was an important reminder of a lesson that many of us had first learned years earlier in our test rover field trials in California. I suspect that military commanders everywhere have known this for a long time: Always look forward, toward your tactical objectives, but don't forget to look back over the whole scene now and then or else you might miss the bigger picture.

View from the Top

Finally, after nearly an Earth year of taking photographs while zigzagging, slipping, trenching, scrambling, and generally rugged climbing, *Spirit* reached the summit of Husband Hill, a broad plateau about 100 meters above the plains where we had landed. While Husband Hill would be considered a rather low hill on Earth, for the rover and the team, reaching the summit was momentous. During the trek we had to survive the depths of winter, deal with a sometimes-flaky right front wheel motor, and figure out how to work and plan efficiently using the relatively small amounts of data that we could return to Earth while Mars was clear across the solar system. Still, the payoff had been huge. We were finally among the kinds of things that we had hoped to encounter—outcrops, layered rocks, physically and chemically weathered rocks and soils—many of which appeared to preserve evidence for the action of liquid water on the surface long ago. The water story is still poorly understood and far from complete in the Columbia Hills, however. From our vantage point high above the plains we acquired our largest Pancam panorama during *Spirit*'s mission, a behemoth 653 image mosaic in five filters covering every square centimeter of terrain from the rover deck out to the horizon. It's a magnificent view. It is a "Rocky" moment as the underdog climbs to the top of the world, and it shows us some entirely new and mysterious places for the first time. Some of these places were hidden from view because they were blocked by those impossibly distant hills that we saw back on SOL one. The terrains beyond the hills, the Inner Basin (to the south) and the East Basin, showed some tantalizing features—rugged, knobby mesas, extremely dark deposits on some hill slopes, and whitish, layered deposits on the upper slopes of McCool Hill in the distance. Perhaps the most interesting and enigmatic feature, though, was a squarish and light-toned patch of ground a few hundred meters wide that we called "Home Plate" based on pictures of it from orbit that show it to be shaped a little like home plate on a baseball diamond.

Spirit crested the summit of Husband Hill right about the time when the available daily solar power peaked as well. In fact, there was more power available than we could routinely use each SOL for photography, other science measurements, or driving, and so the rover was forced to shunt power each day as waste heat. I had the idea of trying to use this otherwise-wasted power to shoot some new nighttime photographs from the rover. Along with astronomer colleagues Mark Lemmon and Mike Wolff, we devised a series of scientifically useful astronomical observations that we could do occasionally at night from what we were now calling the Husband Hill Observatory. We took pictures of the martian moons Phobos and Deimos moving in the night sky, against the same familiar background constellations that we can see at night from Earth. We watched the moons slip in and out of Mars's shadow in "lunar" eclipses. We searched for meteors when Mars passed through the path of Halley's comet, and we basked in the glorious colors of the martian twilight skies. Pancam is a camera, not a telescope, so the images do not have the resolution or feel of telescopic views. But still, the images of curving star trails, potato-shaped moons moving in the night, shooting stars, the setting Sun, and the rising Earth are familiar, yet alien and evocative. These are the things that we could see during twilight and at night with our own eyes if we were on Mars. One day, people really will set up their own backyard observatories on Mars, just like we did for a few enjoyable months with *Spirit* in late 2005.

The inexorable advance of the seasons, however, combined with our south-facing tilt as we descended down the other side of Husband Hill toward Home Plate, meant that we no longer had enough power for routine nighttime observing. On the route downhill, we stopped to photograph and characterize a large field of dark sand dunes called El Dorado. It was poetic that some of our last nighttime observations were looking up at the southern sky constellation Doradus while roaming among the sands of El Dorado. Golden indeed. From back on the summit of Husband Hill, Home Plate appeared to be a small basin with a bright, reddish ring of material on the outer edges of the basin. Some people on the team started calling the bright edge materials a "bathtub ring." Are those some kind of evaporated mineral deposits like sulfates or carbonates or other salts? Or are they hardened dusty coatings like we saw on some of the so-called "white" rocks down in the plains? Or are they something else entirely? Home Plate was first seen a little more than a kilometer away, downslope to the south of the summit. Now it's maybe a few months' drive down to there, or longer if we find some other interesting things on the way, and so we decided that the best way to find out what it really is, is to just go there. Stay tuned.

Sometimes I go back and look at our early pictures from the lander. I can't help but feel like I do when I look at baby pictures of my children. Ah, we were so young, so naive. We were incredibly stingy with our resources, often using fewer color filters than we really wanted or compressing the images to levels a bit harsher than we should have. We were paranoid about the sniper and the fact that every picture, every spectrum, every chemical measurement could be our last. The sniper nicked us once, and the rover almost died a few weeks after landing. If not for the incredible ingenuity and sleuthing of so many clever people working hard together to solve a difficult problem, we would have lost the mission right at the beginning. Since then we've dodged a few more bullets and been on the receiving end of some plain old luck. It's almost impossible to believe that our plucky little machine is sitting there on top of those hills on Mars that are named after a band of fallen colleagues. And we still don't know when our photographic and scientific adventure will end.

Part 3: An *Opportunity* in Meridiani

On July 7, 2003, our second rover, newly named Opportunity, lit up the night skies above Cocoa Beach as its Delta rocket roared off the launchpad and sent the rover toward its own destiny on the windswept plains of Meridiani Planum. *Opportunity* launched nearly a month after *Spirit*, and set the record for the latest launch of a Mars spacecraft within the roughly biannual "launch window" from Earth to Mars. Weather problems and then rocket insulation problems caused the original launch date to slip by a day, then by a week, then by two weeks. As Mars missions go, it was perilously close to being a missed opportunity. But on that clear, moonlit evening, we got off to a picture-perfect and on-course beginning.

No one was taking success for granted when *Opportunity* began her fiery plunge through the martian atmosphere six and a half months later, on the night of January 24, 2004. Even though the team had just accomplished a small miracle three weeks earlier with the perfect landing and deployment of *Spirit*, we were just as anxious as we had been for the first landing. Getting to or landing on Mars is never routine. Nearly half of the spacecraft that have attempted to do so in the last forty years have failed. The tension of *Opportunity*'s impending landing had been amplified by the crisis that had just unfolded half a planet away at Gusev. *Spirit*'s mission had almost ended prematurely with that near-fatal computer software malfunction. But with *Spirit* now on the mend and the focus of the engineers, managers, and scientists now on the safe landing of *Opportunity*, the mood was optimistic.

Second Time's a Charm

As different parts of the team gathered in different buildings and viewing areas at JPL in Pasadena to experience the landing, I felt that old familiar queasiness forming in the pit of my gut. This time, though, it was a little different. We *knew* this landing system could work, we *knew* that the telemetry relays through the MGS and *Odyssey* orbiters could work, and *this* was the "safe" landing site compared to Gusev. It was in the bag. My colleagues and I on the Pancam team, along with a bunch of other science team members, watched the initial events unfold from one of the big sci-

Opportunity's track along the outer rim of Endurance Crater, photographed by Pancam while circumnavigating and looking for a way in. Postcard from SOL 114. (NASA/JPL/CORNELL)

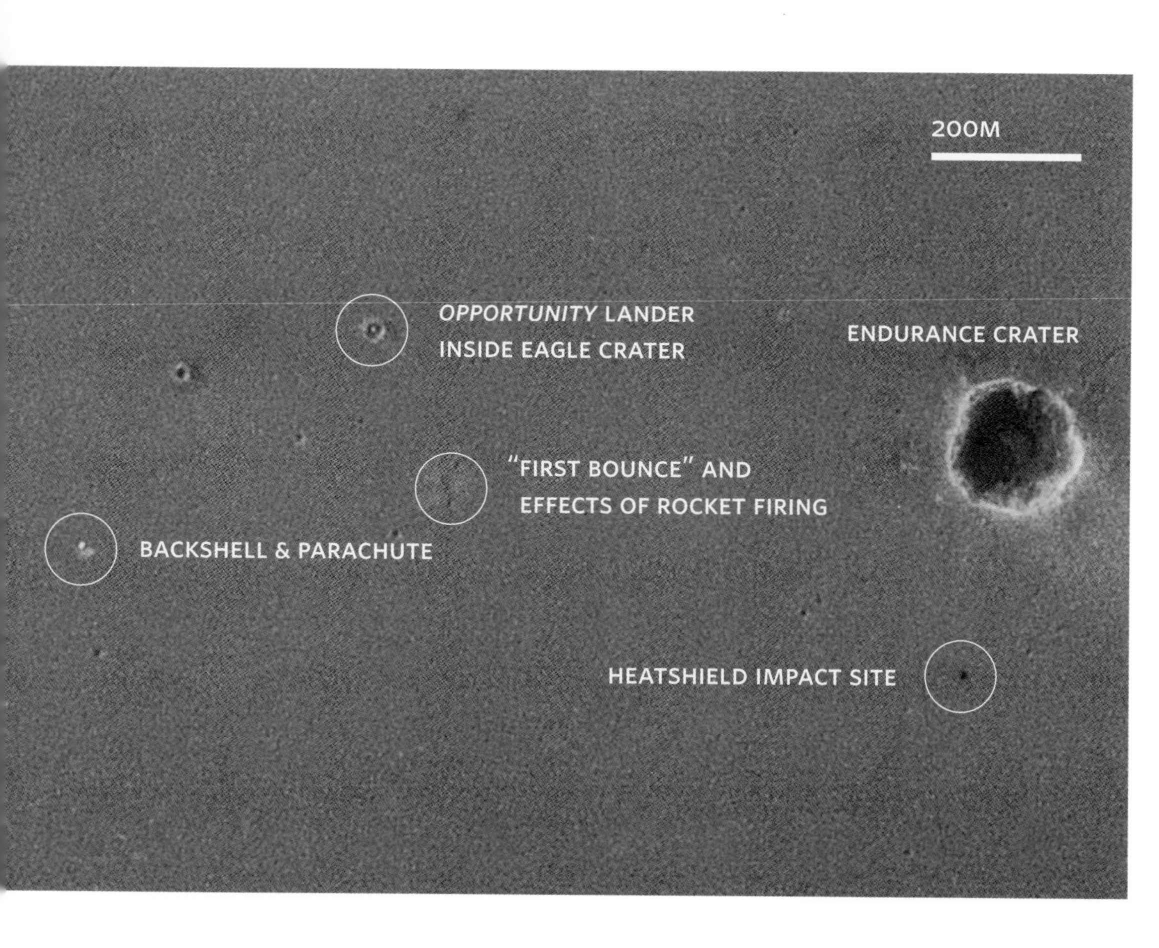
200M
OPPORTUNITY LANDER
INSIDE EAGLE CRATER
ENDURANCE CRATER
"FIRST BOUNCE" AND
EFFECTS OF ROCKET FIRING
BACKSHELL & PARACHUTE
HEATSHIELD IMPACT SITE

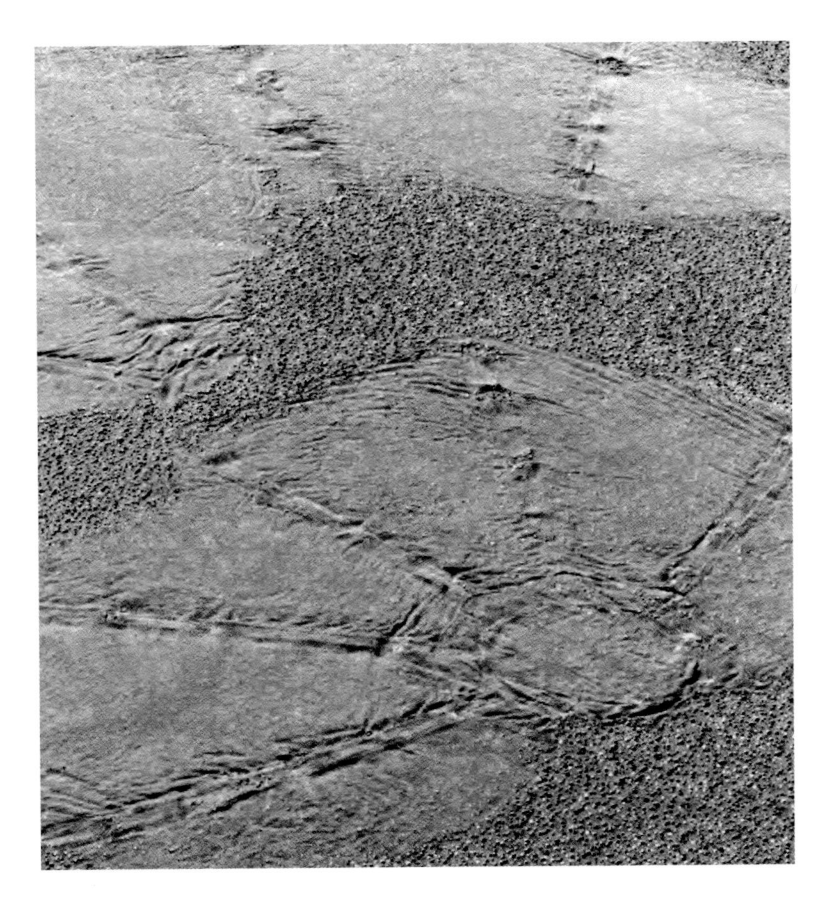

LEFT TO RIGHT
This view of the *Opportunity* landing site is from the *Mars Global Surveyor* orbiter and shows the location of the lander and rover inside tiny Eagle Crater, and the backshell, parachute, and heat shield out in the plains.
(NASA/JPL/MSSS)

Bounce marks left by the airbags inside Eagle Crater appear reddish in this false-color (for mineral analysis) *Opportunity* Pancam view from SOL 2.
(NASA/JPL/CORNELL)

As the rover bounced to a landing, detailed imprints of the airbag seams were preserved in the cohesive soils at Meridiani Planum. *Opportunity* Pancam image, SOL 3.
(NASA/JPL/CORNELL)

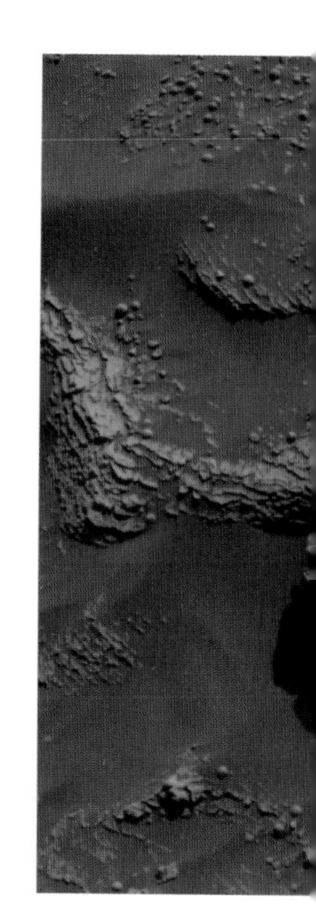

LEFT TO RIGHT
Two *Opportunity* Microscopic Imager views of blueberries. Left: berries on the sandy floor of Eagle Crater (SOL 13). Right: berries embedded in and eroding out of the sedimentary outcrop rocks (SOL 15). (NASA/JPL/USGS)

Opportunity SOL 13 Pancam true-color (left) and infrared false-color (right) views of blueberries and the outcrop rock "Snout" in Eagle Crater. The false-color images help to highlight the locations of blueberries and other deposits. (NASA/JPL/CORNELL)

A RAT hole and "tailings" excavated from the Eagle Crater outcrop rock "McKittrick." *Opportunity* Pancam SOL 36 photo. (NASA/JPL/CORNELL)

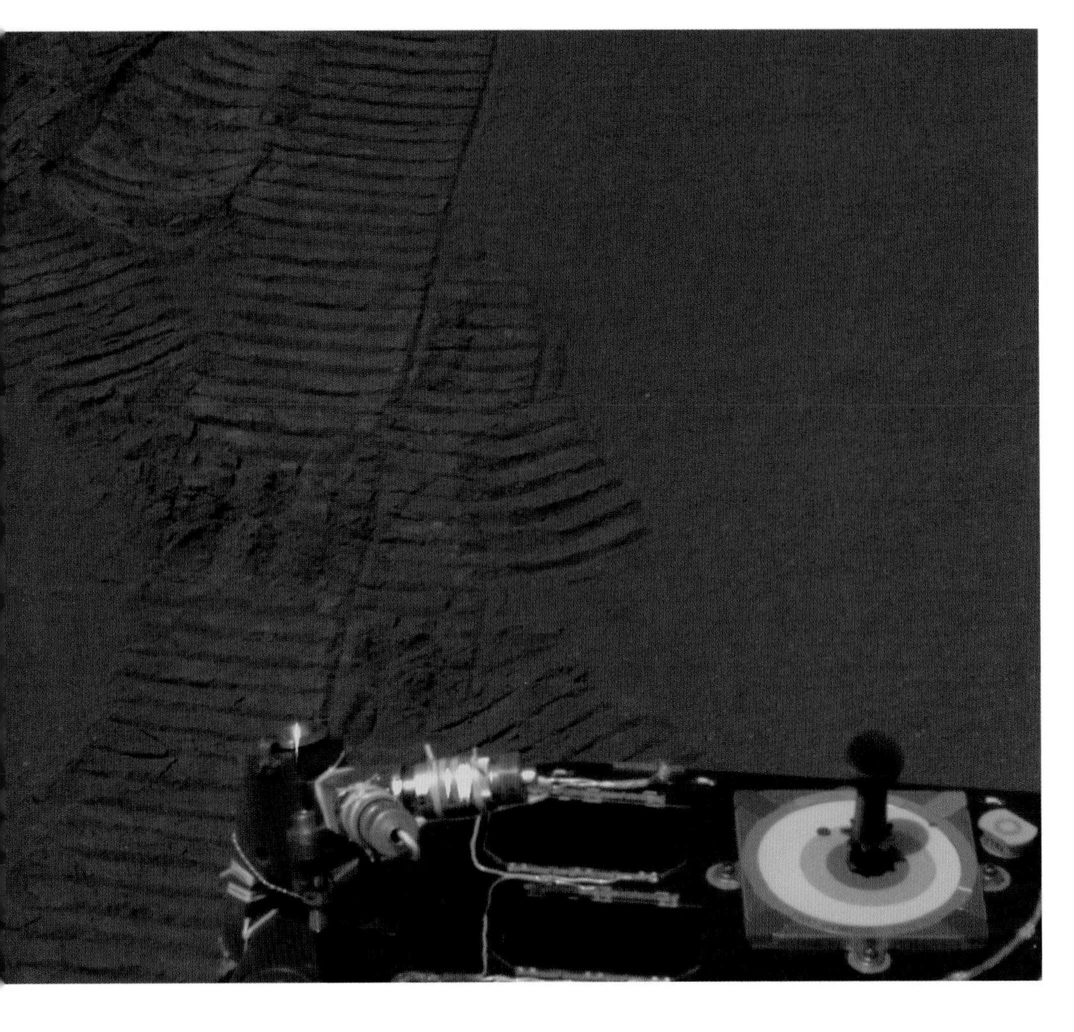

ence workrooms near the main rover mission operations center. This was one of several extremely cool, futuristic-looking rooms that JPL had set up for our use, filled with high-tech computer and communications equipment and ringed with enormous projection screens for computer and video display. The first time I walked into one of these rooms (one of which was known informally as the "Callas Palace," after our JPL Science Manager John Callas) I just had to stop and stare. When I was a kid, *this* is what I imagined the space program should look like.

During the landing, the telemetry showed that *Opportunity* bounced many more times across the plains of Meridiani than *Spirit* did across the plains of Gusev. In retrospect, this was probably because *Opportunity*'s landing site is incredibly flat. In fact, it is one of the flattest places ever explored on the surface of another planet. The first signals didn't give us pictures, but they told us that the landing system had worked and that the rover was basically safe. We were blind to the process of deflating the airbags, unfolding the lander petals, unfolding the rover solar panels, and unstowing the camera mast. These events all happened automatically after the Earth had set below the horizon in the sky above Meridiani. We wouldn't know if we really had a fully functional rover until a few hours later, when the *Odyssey* orbiter would pass overhead and relay the first images and other telemetry back to Earth.

We knew that if things had gone well, the *Odyssey* downlink would contain our first color postcard from Meridiani. The sequence of events that had to literally unfold on the lander was complex. They had been choreographed and tested countless times before in both software and hardware simulations. However, the details of exactly how the various deployments and unstows would occur depended a lot on the tilt of the terrain that we landed on and whether the rover landed right-side-up or not (what the engineers called "base petal down"). If the rover landed upside down inside the lander, the petals were programmed to unfold in such a way as to tip the rover right-side-up again. No one really knew how long all this would take. Some cushion had been built into the timing so that even under the most conservative estimates, it should have finished by the time *Odyssey* passed overhead for that first data relay or "pass." There was some initial imaging from the black-and-white Hazcams and Navcams that was programmed into the sequence, and several of us realized that if everything went smoothly ("nominally," in NASA-speak), then there would probably still be time before *Odyssey* passed over to acquire the color Pancam data as well. So we had programmed the rover to obtain our first color postcard at the very end of what was known as the "sol 1 EDL (Entry, Descent, and Landing) sequence." If things were running slow or there were problems, we'd never get this far into the sequence and we'd have to wait until later to get our first color images. But if everything worked perfectly...

Just before we were expecting that signal from *Odyssey*, most of the science team gathered in the rover mission operations center. Most of the Pancam team was in our small workroom, though, just down the hall. One of the first pieces of information that comes in with a new data downlink from the rovers is a list of what's been done and what's being sent in this particular communications session. Within seconds of receiving the data from that first *Odyssey* pass over *Opportunity* we knew that not only had all of the rover deployments worked perfectly, but that the sequence had run—to completion—all the way through the SOL one postcard! We could see that the images had been taken and that the bits were stored on the rover. We agonized while waiting for them all to make the long journey from Mars to Earth. The speed of light just wasn't fast enough!

The Pancam images were at the end of the queue. We know that they were going to be good because within each downlink we first send tiny 64x64 pixel "thumbnail" versions of all of that sol's images. These thumbnails, the brainchild of JPL rover camera scientist Justin Maki, take up very little bandwidth in the downlink, and they have terrible resolution compared to the full 1024 x 1024 pixel images, but they're good enough to tell us that the data waiting in the flash memory (some of which might not get downlinked in that particular pass) are good. Once the thumbnails were all sent, we watched as the first full-resolution Hazcam and Navcam images came down next. They revealed a strange, alien environment that was almost as different from Gusev Crater as one can get on Mars. The terrain was flat and dark, and there were no rocks like we saw in the *Spirit*, *Pathfinder*, or *Viking* lander images. But wait...Just in front of the rover, maybe ten meters away, a light-toned line of...rocks? Layered rocks? Outcrop? Holy moly!! We didn't know it yet, but we had landed next to the most exciting, scientifically important stuff that had ever been discovered on Mars.

NASA had called a press conference for shortly after the first images had come down from *Opportunity*. The brand-new views of a lovely afternoon in Meridiani Planum would be shared with the world, in near-real time. We knew that the black-and-white images would be showcased in that press conference. However, I knew that the color postcard had been fully acquired and that it, too, would soon be downlinked. I thought it might be possible to wow the world not just with the first images from Meridiani, but with the first *color* images of this strange, exotic place. The timing was tight, but it would be thrilling to have them beamed back and splashed on TV screens all over this planet just a few hours after they were taken on that one.

One by one the pieces of the postcard came in, filling our screens with fantastic high-resolution scenery and filling our heads and hearts with pride and satisfaction that the cameras were working beautifully. The post-midnight JPL press conference was only about twenty minutes away when the last of the postcard came in. We fired up our calibration and mosaicking pipeline to process the images. We'd had three weeks to tweak the software using images from *Spirit*'s Pancams, and so we had a well-oiled machine between the part of our team working at JPL and the many students and support staffers who were working at that ridiculously early hour of the morning back at Cornell University in Ithaca to process and calibrate the images. Just like the landing itself, it all worked beautifully. Onto my laptop computer popped a stunning color postcard of *Opportunity*'s new home. The scene was one of dark, reddish-to-chocolate brown soils sloping away from the lander, and incredibly white, clean (not dusty!) airbags in the foreground. It was beautiful and puzzling and, of course, historic in many ways. The press conference had begun, but after a mad dash across the darkened JPL campus with disk in hand, our photograph was live on television around the world.

Hole in One

Our plan was the same for *Opportunity* as it had been with *Spirit*: Once we landed, one of the first things we'd do was take an initial 360° color panorama to get the lay of the land. Just like on *Spirit*, and on *Mars Pathfinder* before that, this would be called the "Mission Success" panorama for *Opportunity*. In order to declare "success" for our mission, it had been decided that we needed to acquire this particular mosaic, drive the rover off the lander and so many meters around the site, acquire some number of APXS and Mini-TES spectra and other measurements, and survive for ninety sols on the surface.

It has struck me quite often that it is strange to try to quantify and define such an intangible thing as the "success" of a complex, multifaceted endeavor like the rover missions by counting images or meters or spectra and then checking off boxes on some chart. What if we only drove ten meters and the wheels fell off, but we ended up parked right next to a cactus? Is that "failure"? Should we pack it up and go home if we only completed 50 percent of the Mission Success panorama before the cameras died, even if we saw giraffes in the images? These are facetious examples but

PREVIOUS SPREAD (LEFT TO RIGHT)
The dark soils of Eagle Crater and the back of the *Opportunity* rover deck, including the "MarsDial" Pancam calibration target. (NASA/JPL/CORNELL)

Blueberry fields and outcrop rocks on the rim of Eagle Crater. *Opportunity* Pancam SOL 17 false-color view. (NASA/JPL/CORNELL)

An *Opportunity* Pancam SOL 20 close-up view shows blueberries, sand, rocks, and outcrop fragments in a 25-cm-wide piece of the floor of Eagle Crater. (NASA/ JPL/CORNELL)

FOLLOWING SPREADS
124–125
A postcard from the *Opportunity* Pancam SOLS 58 and 60 "Lion King" panorama, acquired from a majestic outlook just outside the rim of 22-meter-wide Eagle Crater. (NASA/JPL/CORNELL)

126–127
The *Opportunity* "Erebus Rim" 360° panorama, including the rover deck. This was the largest Pancam panorama acquired by either rover. It was shot over 43 SOLS (652–694) while problems with the rover arm were being diagnosed. This mega-postcard consists of 1,590 images shot through fourteen of Pancam's filters. (NASA/JPL/CORNELL)

128–129
Opportunity's Rub al Khali ("Empty Quarter") 360° panorama, SOLS 456–464. (NASA/JPL/CORNELL)

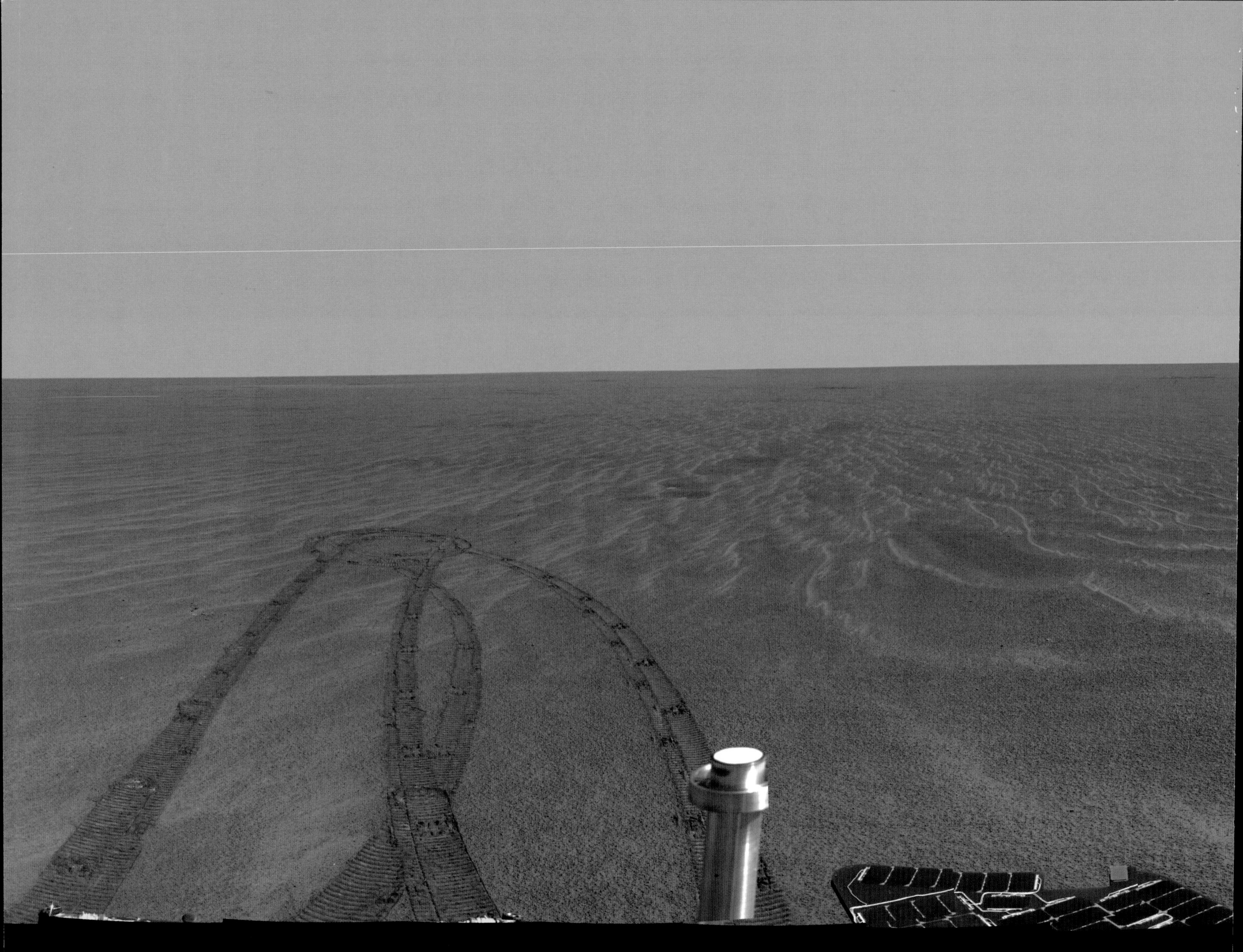

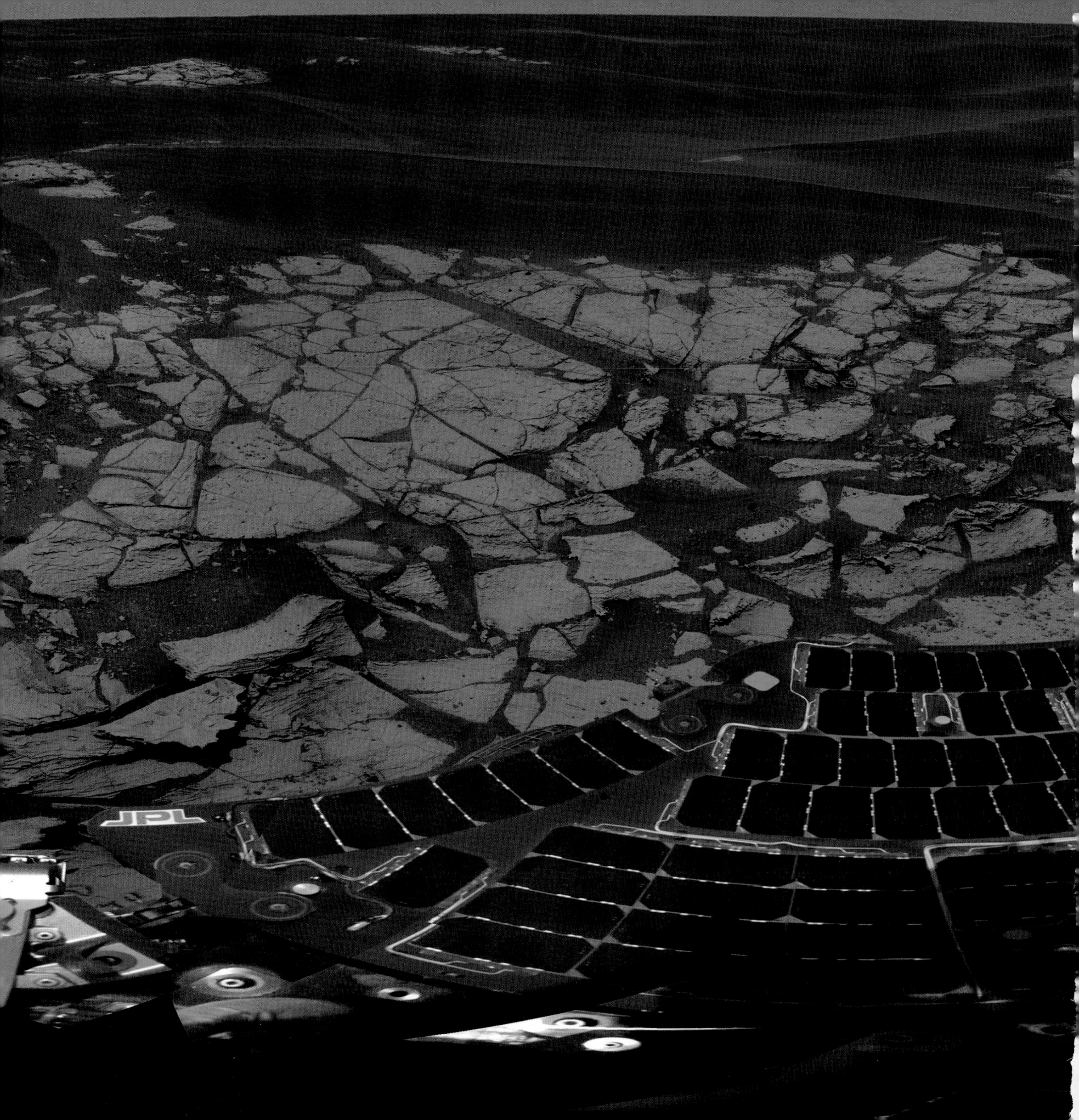
JPL

they underscore the difficulty of trying to micromanage things like the flowing, turbulent, sometimes incremental, sometimes playful human processes of science and exploration. It seemed silly to me to place so much weight on this initial panorama when it was just the opening act in what was hoped to be a long, exciting show.

Regardless of what it was called, the Mission Success panorama was a stunning, revealing glimpse into the unexpected, exotic, and wholly new world that we found ourselves exploring in Meridiani. The first thing many of us realized was that the world that we could see wasn't really flat like we were expecting. There seemed to be high walls completely surrounding us. They turned out to be only a few meters high, but they seemed higher at the time. After a short while, careful stereo imaging revealed that we were actually in a hole, a small hole only about 20 meters wide and 2 to 3 meters deep. We soon figured out that the airbag-laden lander had rolled into and come to rest at the bottom of one of the few small, shallow impact craters in the region. Of all the luck! Golfers on the team liked the name "Eagle" for our tiny little new crater home. It was a long hole in one.

Almost our entire initial worldview of the *Opportunity* landing site came from the bottom of this small crater. Though it was no larger than a typical high school classroom, it made for quite an interesting view. We could see the dark walls of the crater close by, including the indentations left by the airbags as they rolled over the rim of the crater and down into the bowl. We could only see outside the crater in a few small places. The horizon seemed flat out there, but what was it really like? Our attention didn't dwell for too long on the world outside of Eagle, however. The world inside the crater turned out to be too fascinating. To our astonishment and delight, in the very first images sent back we discovered that we were parked right next to a bright band of rocks that was cropping out of the northern half of the crater's inner rim. Unlike anything that we had ever seen before, it was bright, flat, reddish, and layered. Indeed, it was so distinctly layered that we could tell even from 10 meters away with Pancam.

The Mini-TES acquired its own Mission Success panorama of Eagle Crater and found a strong hematite signal in the spectra of the surface. It was the same signal detected years earlier from orbit by Phil Christensen's TES instrument. It was a wonderful vindication of the power of remote sensing to help select a landing site. Phil and I had had some sobering conversations about what it would mean for Mars science if we *didn't* find hematite on the surface at Meridiani. Happily, the discovery of hematite with *Opportunity* became a wonderful demonstration of the value of ground-truth in being able to extend and enhance observations from orbit. This kind of orbital- and local-scale reconnaissance will be critical as scientists try to figure out the details of the history of Mars and how it relates to the history of Earth.

Blueberry Fields Forever

Curiously, in the places where the airbags had packed down the soil, there was no hematite detected by the Mini-TES. How and why did that happen? Where did the hematite go? It was strange, but then it got even stranger. The engineering team had the experience of driving *Spirit* off the lander a few weeks earlier. The surface environment at Meridiani was benign by comparison, and it took the team only seven sols to get *Opportunity* off the lander, onto the surface, and ready to rove. Once we got our new set of six wheels in the dirt, we took a spectacular panorama of the lander and the interior of Eagle Crater, similar to the lander panorama that we had taken for Spirit. NASA had named the landing site the Challenger Memorial Station, in honor of the crew of the shuttle lost in January 1986. That tribute gave the photograph extra emotional charge for many of us.

The first thing we did once we got *Opportunity* onto the martian soil was to take contingency chemical and mineral measurements with the arm instruments. As part of that sequence, we pointed the cameras down for a high-resolution, up-close photo of this strange ground. There are only a few moments on this project when I remember clearly where I was and what I felt when a specific new image first popped up on my screen. Seeing *Spirit*'s first Pancam color image from Gusev Crater was one of those moments. Seeing *Opportunity*'s first close-up Pancam images of the ground in Eagle Crater was another. When we pointed the cameras down and shot the pictures, we discovered that the surface was littered with hundreds of strange and almost perfectly spherical grains. They were just a few millimeters across, about the size of ball bearings or BBs. As we looked around we saw that there were thousands, perhaps millions, of them covering the ground. Most of the grains were reddish and some were grayish, but all of them were less red (or more blue) than the surrounding soils. Someone started calling them "blueberries," and the name stuck. What were these bizarre little things and how did they get there? There was one interesting clue right away: In the "hematite-free" places where the airbags had packed down the soil, we didn't see any blueberries. Had they been crushed and pulverized by the airbags? Or had they been pushed deeper into the soil by the airbags? We had to find out, because the answer might tell us whether or not these strange martian berries were the source of the hematite that had drawn us to Meridiani.

One approach to solving the berry mystery was to take super-close-up pictures of them with the Microscopic Imager. Photographs of the strangely spherical little marbles resting on the sandy ground multiplied. We deployed our two spectrometers onto a patch of berry-filled soil, but it was hard to figure out what the results meant. The berries were so small compared to the measurement region or "footprint" of the instruments. Then we got a break. The Mössbauer spectrometer has a switch on a contact plate that is used to confirm the correct placement of the instrument on a surface. Apparently, when the contact plate pushed against the blueberries, it pushed some of them down into the sand, nearly burying them completely but not crushing them. This provided a clue that the airbags may have similarly pushed the berries beneath the soil in the hematite-free regions seen by the Mini-TES. It was certainly not conclusive evidence, but it supported the idea that the berries were loaded with hematite. We still weren't any closer to figuring out if the hematite had been formed in or by water, however.

Once we finished our contingency soil measurements, we made a beeline for the outcrop. As we got closer, we found that the outcrop rocks were even more interesting than we had thought. We could tell from a distance that the rocks were layered. Now we could see that they were very finely layered. In fact, a geologist would call the rocks "laminated" because the layers are only a few millimeters thick. We could see that some of the layers crosscut one another, and other layers showed gently curved and undulating patterns. We could also see that the rock was pretty porous. This was not at all like the typical volcanic-looking rocks that we were seeing at *Spirit*'s home in Gusev or at other previous landing sites. Huge numbers of blueberries were scattered on and *embedded* in the outcrop rocks. We could even see berries that appeared to be eroding out of the outcrop itself.

Finding the berries inside the outcrop rocks was key to eventually solving the mystery of what they are. Steve Squyres made an analogy to blueberries in a muffin. The "muffin" or porous outcrop rock in this case must be easier to erode away than the blueberries. Over time the muffin part slowly disappears and the blueberries pop out and pile up on the ground. This analogy still didn't explain all the puzzling observations, however.

For more information we turned to the rover arm chemical instruments. The APXS showed a huge amount of sulfur, chlorine, and bromine in the outcrop rocks. These elements are highly mobile and

PREVIOUS SPREAD
Postcard of the small (~8-m-diameter) crater Fram, in the plains of Meridiani Planum between Eagle and Endurance Craters. *Opportunity* Pancam SOL 88 photo.
(NASA/JPL/CORNELL)

RAT hole ground into outcrop rock at Meridiani Planum photographed by *Opportunity*'s Microscopic Imager on SOL 146. Each of the holes on this and the following page is about 4.5 cm across and about 8 MM deep. (NASA/JPL/USGS)

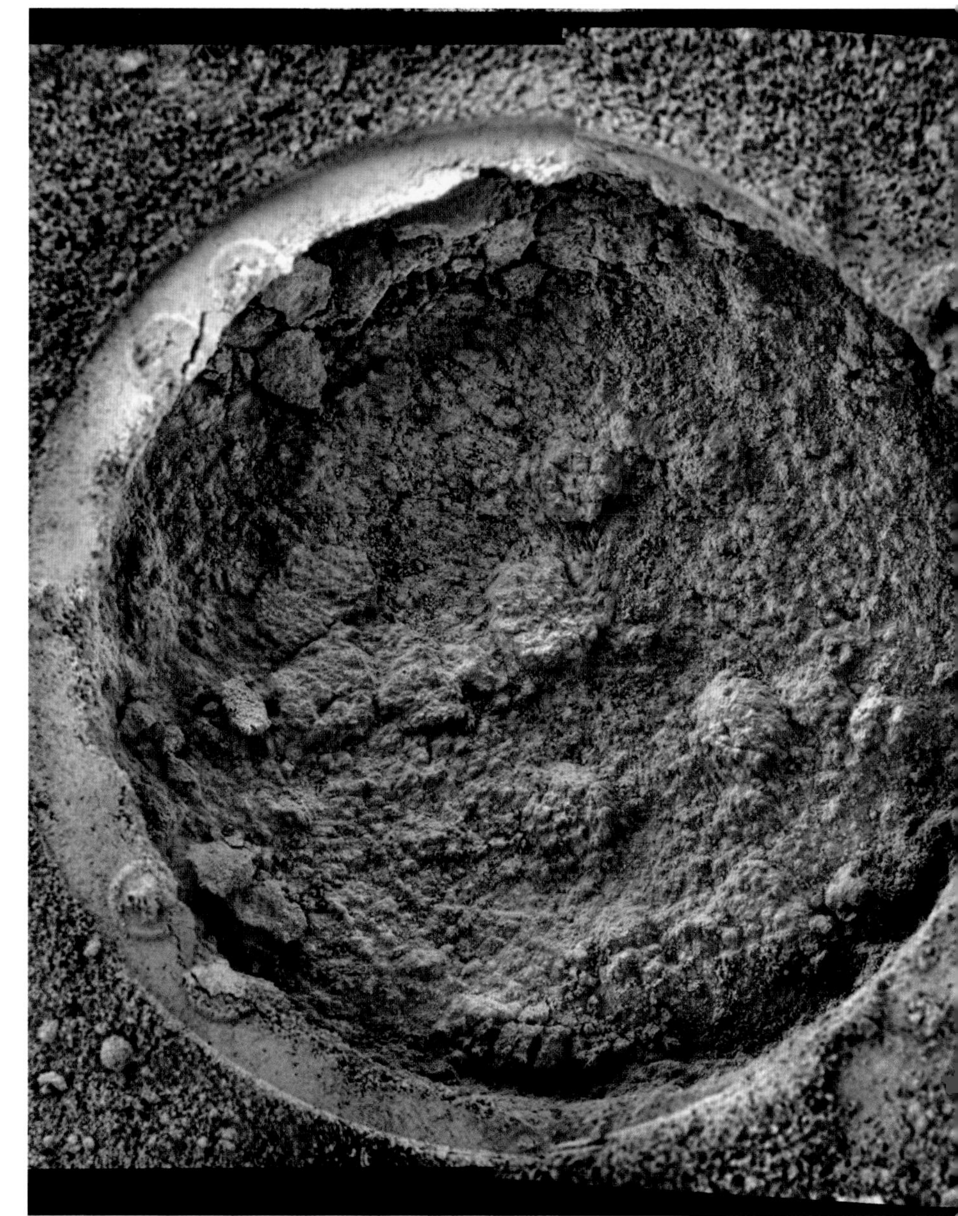

easy to concentrate in some kinds of sedimentary rocks, but they never occur in such high abundances in volcanic rocks. Even more surprising, when we measured the iron minerals in the outcrop with the Mössbauer spectrometer, we found evidence for an exotic mineral called jarosite. Jarosite is an iron-bearing sulfate rock with the chemical formula $(K,Na)Fe_3(SO_4)_2(OH)_6$. Finding a sulfur-rich mineral in such a sulfur-rich rock was not particularly surprising. However, that $(OH)_6$ in the chemical formula was surprising because it meant that this particular mineral was formed in a water-rich environment. Water (H_2O) can be broken into parts by heat and chemical reactions. One such part, (OH), is a common way that water, or at least evidence of water, is trapped in the structure of a mineral like jarosite.

Jarosite had been predicted to occur on Mars by a late colleague and friend named Roger Burns, a brilliant geochemist and planetary science professor from MIT. Roger literally wrote the book on solid state mineral spectroscopy. He knew that there was sulfur in the martian soils, as we all did from the *Viking* lander experiments in the 1970s. He also knew which kinds of minerals that sulfur would occur in on a planet like Mars (or Earth) where there is lots of iron and a possible watery past. It was a real vindication of Roger's work to have discovered this mineral within the outcrop rocks of Meridiani Planum. Many of us on the team who knew and had worked with Roger felt a twinge of regret that he hadn't lived long enough to see his hypothesis verified.

The discovery of jarosite, as well as the enormous sulfur abundances in the outcrop in general, is a source of some concern for planetary scientists thinking about the habitability of Mars for life as we know it. Some of the minerals found in the outcrop appear to have precipitated out of a watery solution as that water was evaporating or seeping underground. The problem is, if you were to dissolve those sulfur-rich minerals back into the water, the result would be a mildly acidic, essentially hydrochloric acid solution. Even a weakly acidic solution can break down organic molecules, or prevent them from forming in the first place. Did Mars have oceans or lakes—that were filled with battery acid? We don't know, but the possibility can't be ruled out with the data that is in hand. If martian water was acidic, what would be the implications for life? There are simple life-forms on Earth, discovered as part of the wonderful "extremophile revolution" in astrobiology over the last decade or so, that live in fairly acidic waters, comparable to water that could reasonably be predicted to have occurred in places like Meridiani long ago. But on Earth, at least, many of those bacterial life-forms are thought to have evolved into such extreme environments from initially more benign ones. Could life—on Earth—actually form in highly acidic environments? We don't know. And because we can't even answer that question on our own home planet, the jury is still out about the possibility of life forming, or evolving, or even still existing, in once potentially habitable places on Mars like Meridiani Planum.

We needed to see into the outcrop rocks to make sure that the elements and minerals we were detecting weren't just from some dusty or dirty surface "contamination." We used the RAT to grind into the outcrop and into berries embedded in the outcrop. However, a curious thing happened: the RAT holes ended up with bright, reddish halos of "drill tailings" around them. We suspected that what was happening was similar to what happens when a student in a college mineralogy class does what's called a "streak test" on a piece of gray hematite. In a streak test, a gray or black, possibly even shiny piece of hematite scratched across a white plate produces a red streak on the plate. The red streak forms because the coarse-grained gray hematite is broken down into finer-grained pieces. These don't absorb as much light as the original grains, so they appear brighter and redder in the streak. We thought the same thing was happening with the RAT, which was streak testing the dark gray blueberries and creating lighter, red dusty deposits around the edges of the hole. It was another piece of evidence that the berries contained hematite. Possibly there was hematite within the brighter, reddish, sulfur-rich outcrop rocks as well.

The clincher came from the Mössbauer (MB) spectrometer measurements of the blueberries. This was a hard problem, though, because the berries are much smaller than the smallest region that can be measured by the MB. Putting the MB down at some random spot might only cover a few berries that might represent only a few percent of the area being measured. Fortunately, we could do much better than a random spot. We found a small rock with a bowl-shaped depression that berries had rolled down into. It was a natural collection of many berries, spaced closely together, and the MB could fit right on top of it. By measuring the spectrum in this so-called "Berry Bowl," then measuring another spectrum in a berry-free part of the same rock, the difference between the two measurements should tell us the composition of the berries. The berries turned out to be loaded with coarse-grained hematite.

It was an important piece of the scientific puzzle that we were trying to solve in Meridiani. Slowly, a story was beginning to emerge. The sulfur-, chlorine-, and bromine-rich, porous outcrop rocks were sedimentary, not volcanic, deposits. The layers meant that the sediments were deposited in some sort of dynamic environment, and the curving, "festooned," crosscutting nature of some of the layers suggested that that environment involved flows in shallow water. The discovery

PREVIOUS SPREADS
136–137 (LEFT TO RIGHT)
Rare, wispy clouds photographed in the skies above Meridiani Planum by the *Opportunity* Navcam on SOL 153. (NASA/JPL/CORNELL/TEXAS A&M/SSI)

Looking back uphill during the traverse down Endurance Crater on SOL 173, Opportunity's Pancam shot this dramatic false-color photo of RAT holes ground into the layered sedimentary rocks. (NASA/JPL/CORNELL)

Thin fracture-filling materials photographed in outcrop rock in Endurance Crater. *Opportunity* Pancam SOL 170 false-color photo of a region about 20 cm across. (NASA/JPL/CORNELL)

Pancam false-color closeup view of RAT holes in Endurance Crater outcrop rocks. *Opportunity* SOL 150. (NASA/JPL/CORNELL)

High altitude cirrus-like clouds photographed by the *Opportunity* Navcam on SOL 290. (NASA/JPL/CORNELL/TEXAS A&M/SSI)

LEFT
Eerie dunes and sandy tendrils on the floor of Endurance Crater, *Opportunity* SOL 187 false-color (for mineral analysis) Pancam photo. (NASA/JPL/CORNELL)

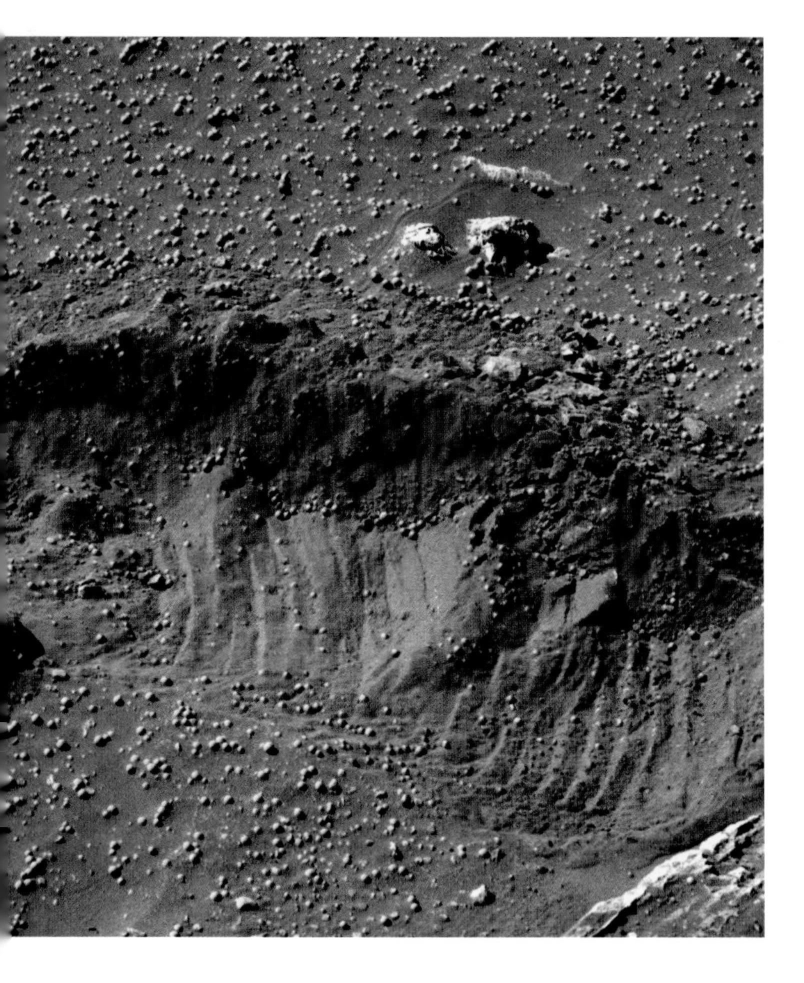

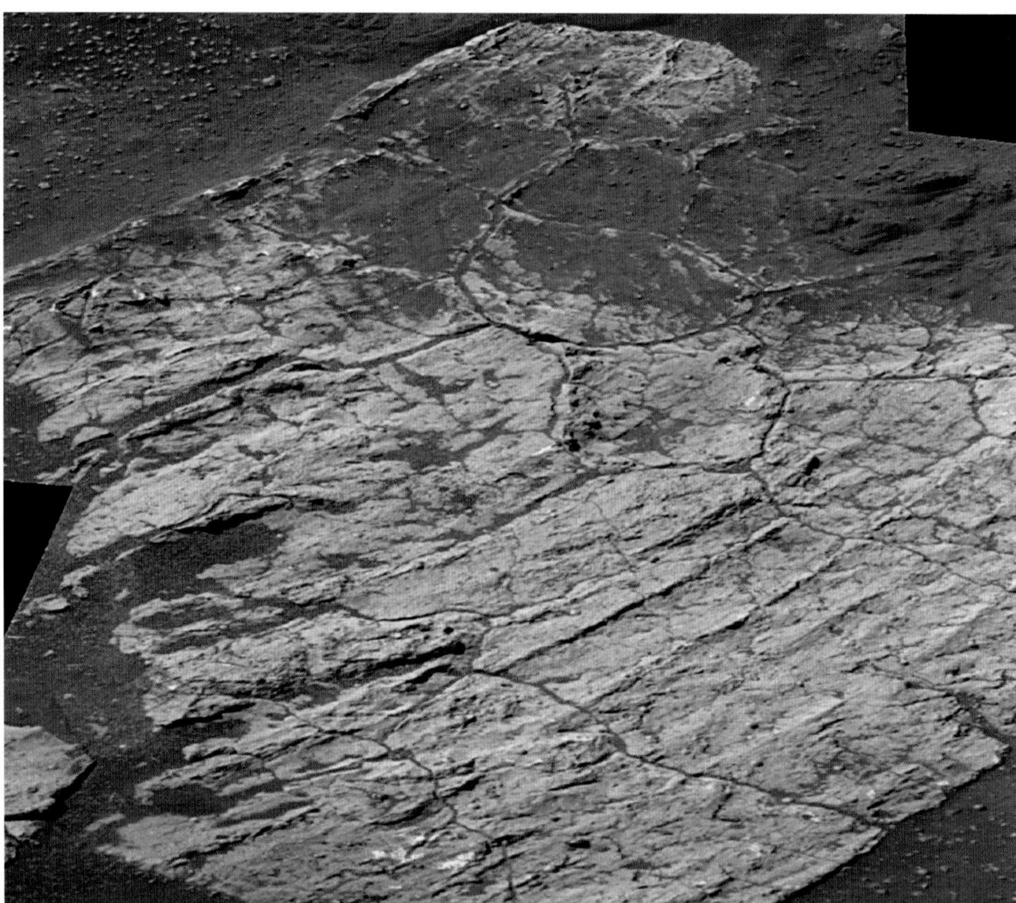

of jarosite in the outcrop meant that there was liquid water involved in its formation. This was supported by the discovery of coarse-grained hematite in the blueberries.

The berries themselves, we hypothesized, were a kind of mineral deposit called a concretion. Concretions are iron-rich spherical rocks found on Earth in places where water completely saturates porous, sedimentary rocks. As the water slowly evaporates or recedes, iron-bearing minerals slowly precipitate out of the watery solution. If the mineral growth is slow enough, the mineral deposits grow in perfect little spheres, which remain embedded in the pore spaces of the rock after all the water is gone. On Earth, concretions can grow to be the size of marbles or even golf balls. Where we landed at Meridiani, for some reason, they only got to be the size of BBs or small ball bearings.

Apparently, the outcrop rocks and the blueberries are the deposits left over after the evaporation of a salty body of liquid water. Here it was—right in front of us—an extensive deposit of layered sedimentary rocks formed in liquid water on Mars. Within a few weeks of landing, we had found key evidence that there was once liquid water on Mars and that the environment must have been much more Earthlike at some point in the distant past. We don't know how much water was there, or how deep it really was, or exactly when it was there, or for how long. But we now *know* that it was there, either on the surface in a lake or shallow sea, or just below the surface in extensive underground aquifers or groundwater systems. A long time ago Mars may have looked much more familiar to us than it does now.

Flatland

We spent nearly two months driving *Opportunity* around inside tiny little Eagle Crater. We took pictures and infrared spectra. We made detailed chemical and mineral measurements of the outcrop rocks and the soils. We drilled more RAT holes and dug some trenches with the rover wheels to look at the subsurface. Ken Herkenhoff's team at the USGS even planned tiny little "micro panoramas" by scanning the Microscopic Imager across some of the finely layered rocks. Exploring Eagle Crater took the combined talents of hard-rock geologists, sedimentary geologists, geochemists, and remote sensing experts, all working closely with the rover engineers, mission planners, and instrument teams. By the time the decision was made to leave Eagle and head out onto the plains around sol 56, we had thoroughly photographed and documented most of our little hole in one.

It took two tries to get *Opportunity* out of the cup. On the first try the rover got bogged down in the sandy slopes on the inside of the crater wall and slipped too much to get out. On the second try, at a different part of the crater wall, there was still some slippage but we finally made it out. We turned around to take a panoramic look with the cameras back at Eagle Crater from a point just outside the rim. One of the team members with young kids likened the situation of the rover looking out over the crater from the rim to the scene in the Disney *Lion King* movie where the wise old baboon is perched atop a majestic outlook, presenting the young lion cub to the gathered animals below. So, the photograph became known as the "Lion King" panorama.

LEFT TO RIGHT

Part of a 50-cm-long trench dug into blueberry-rich soils inside Endurance Crater by *Opportunity*'s wheels, SOL 208 Pancam false-color (for mineral analysis) photo. (NASA/JPL/CORNELL)

A Pancam SOL 208 postcard of the rock called "Escher," which shows possible evidence of mud cracks. This rock was near the lower limit of *Opportunity*'s traverse into Endurance Crater. (NASA/JPL/CORNELL)

Dusty sand dunes and outcrop rocks near Erebus Crater. *Opportunity* Pancam SOL 644 false-color photo. (NASA/JPL/CORNELL)

Once we were out onto the plains, the magnitude of the vast, lonely *flatness* of the new world in which we had landed sunk in. There were few landmarks to use to gauge distance or heading. Eagle crater was one landmark, and again we were struck by how amazingly lucky we had been to roll right into that small hole with our airbags. We could see what looked like shiny metallic or white debris off in a few directions. Close inspection of certain photographs revealed those to be the lander's parachute and backshell about 500 meters away to the southwest, and the lander's heat shield about a kilometer away to the southeast. The only other feature that we could see on the horizon was the rim of a much larger impact crater, about 700 meters to the east. We could tell from MGS and *Odyssey* orbital images and the descent images that the rover had acquired that this crater was about seven times the size of Eagle Crater, or close to 140 meters across. The eastern (far) rim of the crater was higher than the western (near) rim, allowing us to see what looked like brighter, possibly layered materials exposed on part of the crater's inner wall. This pushed the limit of resolution of the Pancams, however, so we couldn't be sure of exactly what we were seeing. As the Sun set in the west every day and brightly lit up that west-facing inner wall, it seemed to be beckoning us.

There were other reasons that the crater to the east should be our next exploration site. If there was layered outcrop exposed near the surface there, like what we found at Eagle Crater, then that would indicate that the water we believed was present where Eagle is at one time also had extended

across at least 700 meters away, and perhaps farther. It was important to find out if the deposits in Eagle were just some crazy little local oddities, or if they were representative of processes that must have acted across a wide region on this part of the planet. More importantly, this new crater was wider, so it was probably also deeper than Eagle. Given that it takes sediments significant amounts of time to build up, layer upon layer, if we could find and explore deeper layers exposed inside the walls or on the floor of the crater, we could be exploring deeper and deeper back in time. In a sense, if we were fortunate, it could be just like hiking farther down into the Grand Canyon to get to older and older rocks exposed in those differently colored layers.

To get to that crater would be a challenge, though, just like getting to the Columbia Hills was for *Spirit*. The rim of the crater was farther away than the 500 or 600 meters that the engineers had estimated that the rover would be capable of covering. We had already chewed into that total with our two months of scrambling around inside Eagle Crater. Realistically, just like for *Spirit*, heading toward the most interesting feature around was really the only choice that we had. Partly because we would have to endure a long drive to get there, and partly to honor the expedition of South Pole explorer Sir Ernest Shackleton and his crew, we named the crater "Endurance" after Shackleton's ship. Naming craters in Meridiani after famous ships of exploration became our new theme. Eagle, after all, was the name of the famous ship piloted by Neil Armstrong and Buzz Aldrin during *Apollo* 11's first landing on the Moon.

LEFT TO RIGHT
Sand ripples and small rocky fragments in the plains of Meridiani, *Opportunity* Pancam SOL 528 false-color (for mineral analysis) photo. (NASA/JPL/CORNELL)

False-color Pancam postcard of sand ripples in the plains south of the heat shield, *Opportunity* SOL 447. (NASA/JPL/CORNELL)

Opportunity Microscopic Imager photo of blueberries near the Mössbauer spectrometer's imprint in the sands of Endurance Crater on SOL 235. The field of view here is about 3 cm across. (NASA/JPL/USGS)

Dark sand and rocky outcrop along the traverse on the inside of Endurance Crater, *Opportunity* SOL 237. (NASA/JPL/CORNELL)

Desolation among the sandy ripples of Meridiani Planum, *Opportunity* Pancam SOL 385 postcard. (NASA/JPL/CORNELL)

Frost (white patches) on the ground and on the MarsDial during the predawn hours of SOL 257. (NASA/JPL/CORNELL)

We used the orbital and descent pictures to map out our route to Endurance. Small craters and other features that we could see in the images became waypoints or buoys to stop at along the way. The flat plains of Meridiani, just like the floor of Eagle Crater, were littered with millions of blueberries, covering the ground and stretching out as far as our eyes could see. Amazingly, we could see only one large rock out on the plains in the route planning images that we were acquiring. This certainly made the job of deciding which rock to study a lot easier than it was on *Spirit*. The rock was not too far from the rim of Eagle Crater. We called it "Bounce Rock" because it appeared that the airbag-laden lander had bounced right on top of it, possibly even breaking a piece off, as it was rolling across the plains. What are the odds of hitting the only rock for hundreds of meters around with the airbags? we wondered.

Bounce Rock turned out to be a piece of volcanic rock with a chemistry very different from the outcrop rocks, blueberries, and sand deposits of Eagle Crater. However, its composition is very similar to some of the meteorites found on Earth that are known to have come from Mars, like the famous ALH 84001. What is a single lonely volcanic rock like this doing so out of place? We think that it is probably an interloper, tossed far across the plains by some distant impact cratering event.

As we continued east toward Endurance, we stopped to study some enigmatic cracks or fissures in the ground that we could just barely see from orbit. These might be chains of small impact craters, or they could be some kind of small rifts created by tectonic forces in the region—we weren't sure.

What we could tell, though, was that there were layered sedimentary outcrop rocks just like the ones we saw at Eagle Crater exposed within only a meter or so of the blueberry-rich, sandy surface.

Without all the rocks and sand dunes that *Spirit* constantly had to avoid or gingerly drive over, *Opportunity* was routinely able to drive 50 to 100 meters a sol. Some days our drive distance was limited by power availability, and other days it was limited by our own curiosity. For example, our next long stop was at a small crater called Fram, only about a third the size of Eagle Crater. Again, we saw the sulfur-, chlorine-, bromine-rich outcrop rocks like we saw at Eagle wherever the subsurface had been dug into and exposed by small asteroid or comet impacts. In fact, for the duration of the one-month traverse to Endurance, we continued to see outcrop rocks everywhere that the surface was disturbed in some way. Finding all of this outcrop rock confirmed that the water-formed materials are spread across this entire region. That provided support for the idea that long ago there was a lot of surface water in this part of the planet.

LEFT TO RIGHT
Six months separate the bright *Opportunity* rover tracks at right (made when the rover entered Endurance Crater) from the newer dark tracks at left (made when the rover drove out of the crater). Tracks appear to brighten with time by trapping brighter windblown dust. SOL 317 Navcam photo. (NASA/JPL/CORNELL)

Layers and "crossbeds" in the walls of Endurance Crater near Burns Cliff, *Opportunity* Pancam SOL 288 "super resolution" image. (NASA/JPL/CORNELL)

Endurance Rewarded

Opportunity finally arrived at Endurance Crater in late April 2004, at the end of her nominal ninety-sol warranty. The traverse to the east had been just as much of an achievement for robotics as *Spirit*'s initial traverse to the north to Bonneville Crater had been, but that's pretty much where the similarity ended. Reaching the rim of Bonneville Crater with *Spirit* was bittersweet. We didn't find any of the outcrop rocks or possible water-related deposits that we had hoped for when we first peered into that big, dusty, sandy hole in the ground. Reaching the rim of Endurance Crater with *Opportunity* was just plain sweet. Endurance is a hole in the ground about the size of a typical football stadium (including the seats). It had everything that Bonneville didn't. The rim and upper parts of the walls exposed large deposits of flat, layered, sedimentary outcrop rocks like those that we saw at Eagle and all across the plains. Below those upper layers, exposed as in some kind of mineralogist's dream, we could see layer upon layer of outcrop rock. Some of the layers had a slightly different color from others. There were many different textures seen in the crater walls. About the only thing Bonneville-like was that, at the bottom of Endurance, about 20 to 25 meters deeper than the rim, there was an enormous field of dark sand dunes. The dune field was spectacular—about 50 meters across, with some individual dunes taller than a meter or two from crest to trough. Though it was beautiful, we knew that it was essentially a big rover death trap that we might be able to drive into if we tried, but could probably never get out of.

Opportunity spent a few weeks circumnavigating much of the crater's rim, and we photographed several large and spectacular color panoramas of the crater's interior. We were also making stereo measurements required to estimate the slopes along the inner walls, trying to find out if there was a way to drive into the crater to get at these deeper layers of outcrop rocks. There were places along the rim that dropped off steeply at 60° to 90° angles, and so the rover drivers had to be especially cautious. We wanted to get close to the edge to get the best views in, but if we got too close and fell in, that would be the end of the mission. Fortunately, there were places along the eastern wall where the slopes were gentler, only about 20° to 30° in places. We could drive in there, but it would be tricky. One of the rover drivers commented that going down into Endurance with all those slabby rocks and blueberries would be like driving on plywood covered in small marbles. The engineering team back on Earth started doing a bunch of specially designed tests using one of the spare/test rovers at the JPL Mars Yard in Pasadena. They drove the rover up and down steeply sloped and slippery surfaces designed to simulate the drive down into Endurance. Again and again the rover proved its ability to handle even 30° slopes. In fact, the team found that the rover could drive over these slippery surfaces better than the team could walk over them. Sometimes, having six cleated wheels can be an advantage.

We found a place to drive in, named Karatepe (by team member, top-notch sedimentary geologist, and fellow Red Sox fan John Grotzinger) after a geologically and archeologically important national park in Turkey. The rover drivers first did a series of "toe dips" by driving *Opportunity* a few meters down into the crater, then hitting the brakes, shifting into reverse, and coming back out. I can only imagine what that hypothetical martian, strategically watching from behind some nearby rock, would have thought about the rover's apparent fits and starts, in and then out ("Make up your mind!"). Well, the toe dips worked and proved that we had traction on these deposits, so we decided to jump in, carefully and methodically, working our way down into Endurance.

It was a snail's pace crawl down into the crater that took months. As we drove down, we would take some photos of the region in front of us and use the pictures to distinguish different layers in terms of their colors or textures. Then we would place the arm instruments down on the different layers to make chemical and mineral measurements and to take Microscopic Imager pictures of the undisturbed layers. Then we would either brush or grind into the layers (or both), taking more arm instrument measurements at each step of the process. We'd finish work in one layer, then creep downslope to the next. It was slow and meticulous work, a bit frustrating for some folks on the team who wanted to charge on down the road, but was the kind of careful sedimentary deposit traverse campaign that most of us would have done ourselves if we had been there out in the field.

As we descended, we ground holes into the outcrop and made Swiss cheese out of this section of the crater wall. We were primarily looking to find whether the sulfur-rich, blueberry-rich outcrop rocks continued as we went deeper into the subsurface. They did, and we eventually found that they continued down as far as we could go into the crater—down nearly 10 meters deep to where it became too sandy to drive farther. We were also looking for trends in the chemistry, mineralogy, or texture as we went deeper into these presumably ancient rocks. Certain kinds of trends are often seen in similar sedimentary rock sections on Earth, and so they might help to provide some clues about how these rocks were formed and deposited. This kind of information could possibly even tell us how long all of this took.

We did see trends. For example, we saw both sulfur and magnesium increase or decrease slightly by the same amount in the different layers, but chlorine steadily increased dramatically as we got deeper. This argued that there is some kind of magnesium-sulfate salt mineral that is common among the outcrop rocks, and perhaps differences in the water-to-rock ratio during the formation of these rocks could explain the variations in chlorine, which is a very mobile element in the presence of water. We saw trends in the texture, too. For example, near the deepest parts of the crater that we could access, we discovered a different class of blueberries. These were rougher and covered by what looked like cake batter, making them look more like popcorn than blueberries.

Perhaps the most interesting textures that we saw on the Karatepe drive were a series of distinctive cracks in some rocks closer to the bottom of the crater. The best examples, seen in a rock named Escher after the famous artist and illustrator, were cracks arranged in polygon shapes with raised rims along the edges of the polygons. When the first images of Escher came down, we had another of those special moments when gasps went up in the room. There was a short lull, and several people, at once, said, "Mud cracks!" Mud cracks form on Earth in places where wet, porous rocks (or soils) slowly dry up. As the water dries up, the volume of the rock or soil shrinks and what are known as desiccation cracks form, often in polygonal patterns and often with raised rims along the edges of the polygons.

It was hard to resist the analogy, but we had to acknowledge that there are other possible ways besides drying up a puddle of water that similar cracks can form. For example, stresses related to the impact that formed Endurance, or freeze/thaw expansion from occasional small amounts of surface or subsurface ice, could have formed similar patterns. Interestingly, the cracks in Escher are *superimposed* upon the layers in the outcrop rock. This indicated that whatever the origin of the cracks, it was an event that occurred after the original sediments were laid down, perhaps even after the impact that created Endurance. Did Endurance fill with water one or more times after it formed? Could this water have led to the formation of the popcorn-blueberries, or the chemical trends seen in the Karatepe section? Could those cracks in Escher have formed when that water dried up? We still don't understand the reasons for many of the trends in chemistry and texture that we saw during the nearly four months that *Opportunity* spent driving down, but they will certainly be the topic of intense research by team members and others over the coming years.

When we got down into the outer fringes of the huge dune deposits at the bottom of Endurance Crater, the rover drivers and mission managers decided that we couldn't go any farther. This was disappointing to many people on the team. Sand is everywhere on Mars, and the movement of sand by the wind is one of the most dynamic geologic processes occurring on the planet today.

FOLLOWING SPREADS

150–151
Opportunity's Pancam SOL 322 "self-portrait" postcard. By this point in its martian journey, *Opportunity* was much less dusty than her sister *Spirit*. (NASA/JPL/CORNELL)

152–153
Opportunity Pancam SOL 393 postcard of "Naturaliste" crater, in the plains south of Endurance Crater. (NASA/JPL/CORNELL)

154–155
Postcard of the heat shield impact site, south of Endurance Crater. This is the youngest (known) impact crater on Mars. *Opportunity* SOL 330. (NASA/JPL/CORNELL)

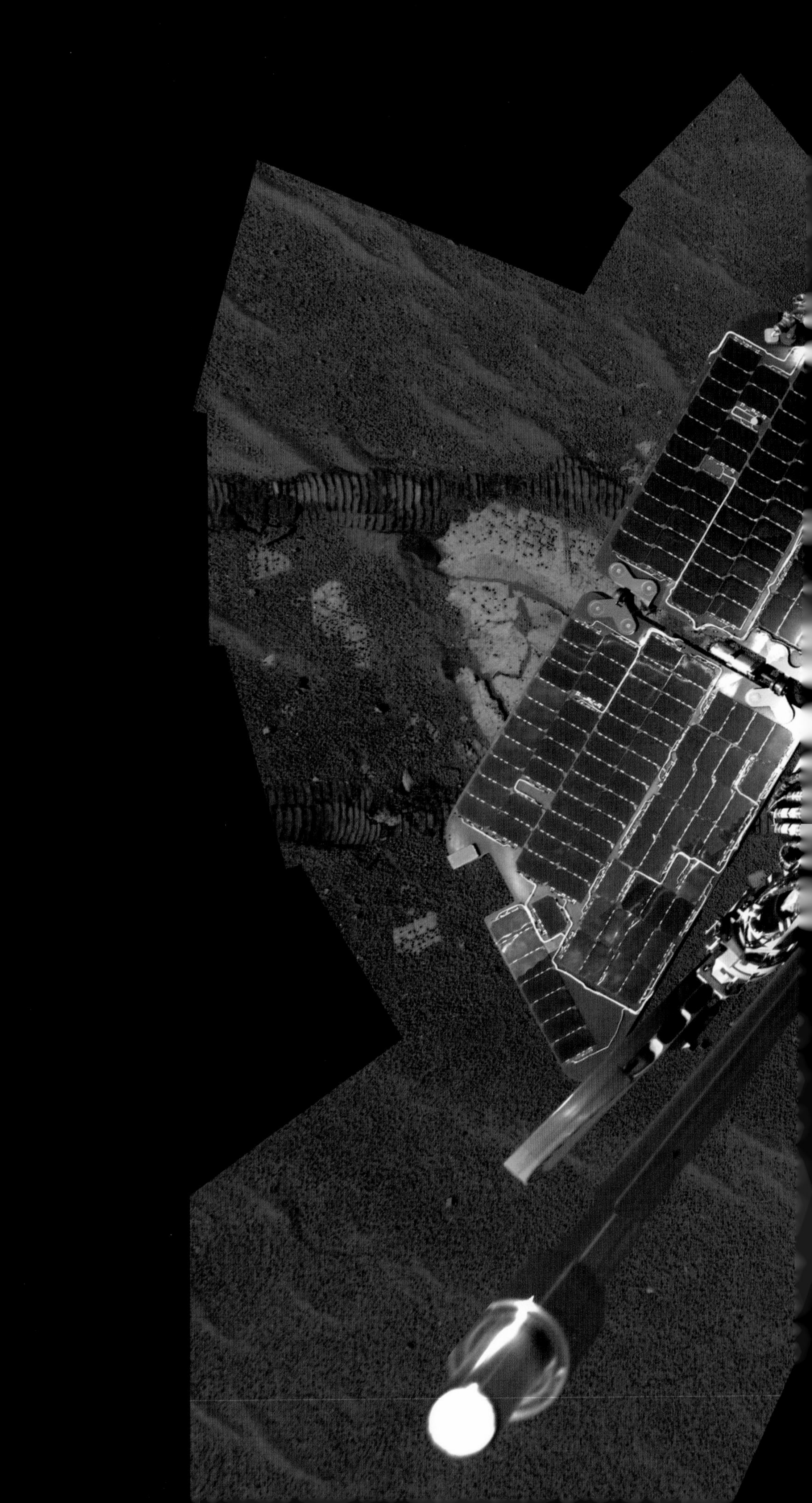

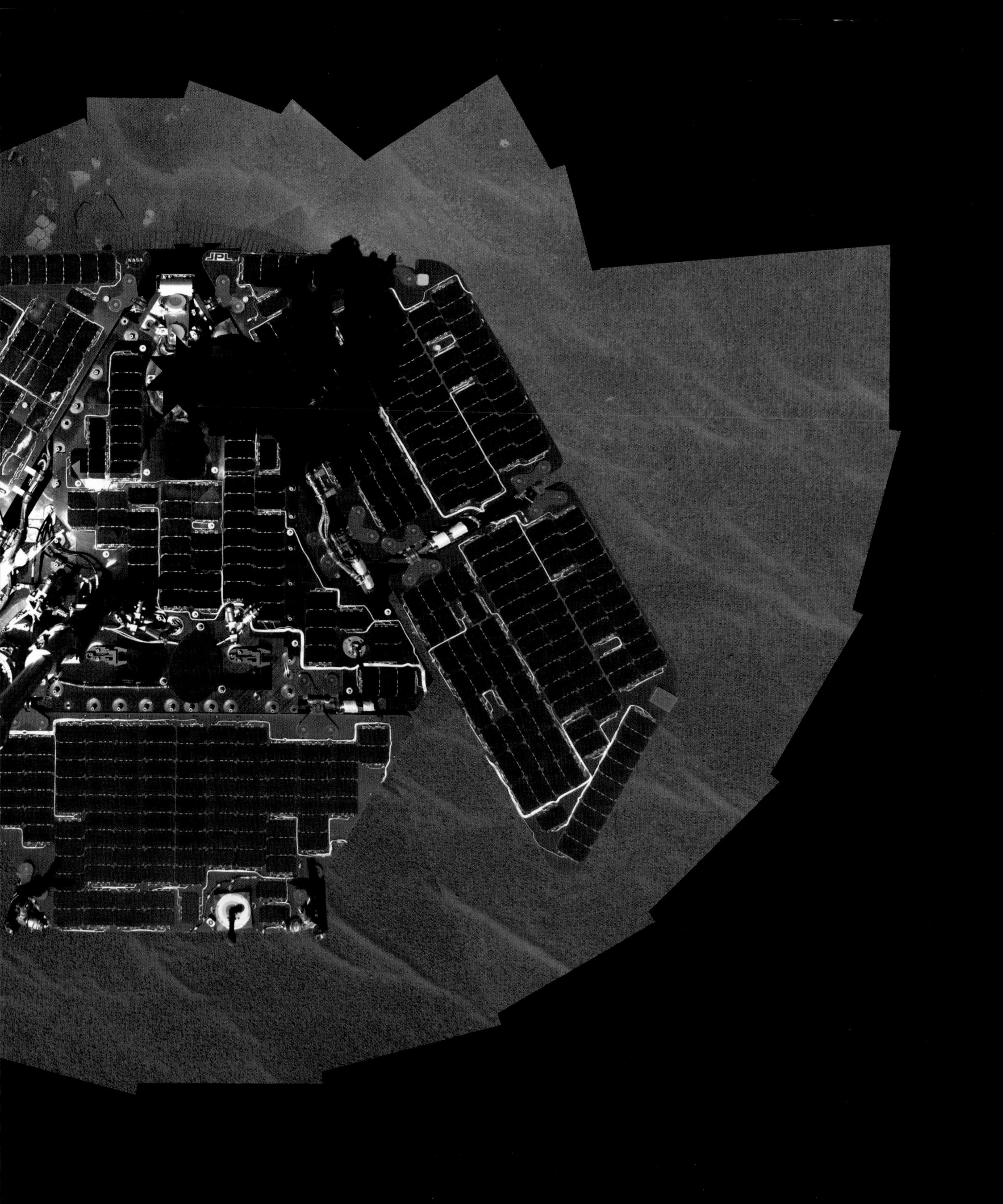
NASA
JPL

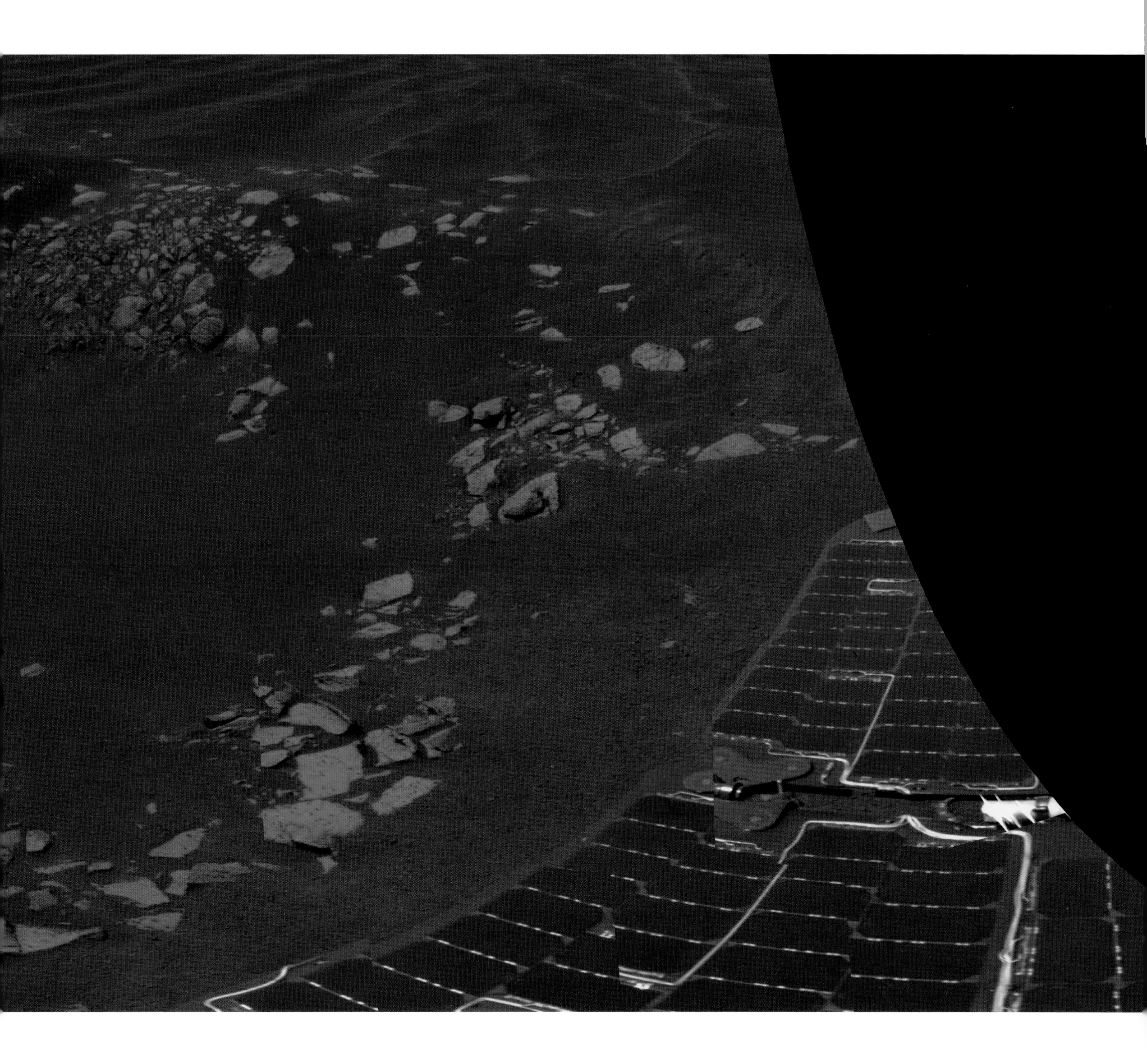

A 20-cm-wide iron-nickel meteorite, discovered near the heat shield impact site, *Opportunity* Pancam SOL 352 photo. (NASA/JPL/CORNELL)

Opportunity Pancam SOL 580 false-color (for mineral analysis) photo of outcrop rocks and rinds near Erebus Crater. (NASA/JPL/CORNELL)

Blueberries, sand, and wheel tracks in part of a trench south of Endurance Crater, photographed by the *Opportunity* Pancam on SOL 373. (NASA/JPL/CORNELL)

RAT holes ground into rinds in sedimentary outcrop rocks north of Erebus Crater, Meridiani Planum. *Opportunity* Pancam SOL 561 false-color (for mineral analysis) photo. (NASA/JPL/CORNELL)

Even more poignantly, sand dunes like these at the bottoms of craters are some of the most common features on Mars. Here we were, parked right next to a typical, lovely martian sand dune. If we could make some measurements of this sand with our instruments, we could learn new things about one of the dominant landforms on the planet, essentially learning new things about other places by extending our knowledge outward from this one. We could take pictures of it, of course, using many different color filters and at many different lighting angles. These kinds of "photometry" experiments were often conducted by team members Jeff Johnson, Ray Arvidson, and others as a way to remotely probe the physical properties (sand grain sizes, packing density, etc.) of the surface materials. We could also acquire Mini-TES infrared spectra and temperature measurements of the dune field. But we really wanted to touch it, to measure its chemistry and mineralogy directly, and to dig into it with the wheels. Just a little...But we were told no, it's too risky, and so we drove away. It was a frustrating reminder that in the rover business, just as with all space missions, the health and safety of the spacecraft (and crew) come first.

Burns Cliff

We took our safe and healthy rover and headed back upslope, away from the sand and toward some spectacular layered deposits seen along the northern wall. These layers were near some of the steepest slopes inside the crater, near a place that we had dubbed "Burns Cliff." By angling along the inside of the bowl and slowly heading uphill, we could keep the rover at an acceptable tilt. The bonus was that by driving along at that tilt on that side of the crater, we could keep the solar panels pointed more directly into the sunlight, improving our power situation. Fortunately, the power decrease for *Opportunity* was not as bad as it was for *Spirit* as the Mars seasons progressed into winter. *Opportunity* is much closer to the equator, so the Sun did not get as low in the sky at Meridiani as it did at Gusev. Still, we had to be stingy. Partly this was because of a handicap that *Opportunity* had that *Spirit* didn't. Shortly after landing, back in January 2004, the rover engineers noticed that one of the heaters on the robotic arm had become stuck in the "on" position. Apparently, the only way to shut the heater off was to shut down many higher-level systems. Doing this would also shut off other heaters, the most important of which was a survival heater for the Mini-TES instrument. If the Mini-TES heater was shut off, there was a chance that some of its critical internal optical components would get too cold and crack, thus rendering the instrument useless. So for most of the mission, we simply had to live with losing a lot of power every night because of the stuck heater. There was no other choice because Mini-TES was too important a piece of the payload to lose.

As it got closer to winter solstice, however, the amount of power that was being lost because of the stuck heater was becoming a greater and greater percentage of each sol's supply. If we kept going at this rate, we'd eventually spend all our power on the stuck heater and wouldn't have any left for driving or picture-taking. The engineering team came up with a workaround called "deep sleep" that would shut off the stuck heater and other systems overnight and would save us a lot of power, but at the risk of possibly killing Mini-TES. It was a tough call. Steve Squyres, Phil Christensen, his instrument engineer Greg Mehall, and many of us on the science team anguished over the decision. In the end, there really wasn't much choice. Mini-TES had to risk dying or the entire mission would be at risk. So we started deep sleeping, and as predicted, our available power went up significantly. Somehow, magically, Mini-TES kept working. Would our luck ever run out?

Burns Cliff was one of the most photogenic and precarious places that *Opportunity* had ever examined. I imagine a helicopter view of our rover parked sideways at nearly a 30° slope on the inside rim of a hole in the ground the size of the Rose Bowl Stadium. The rover would be carefully scooting forward, slipping downhill a bit, then scooting forward a bit more to get its arm instruments deployed on some of the magnificently exposed layers of rock in the crater wall.

The results from the pictures and chemical measurements were just as breathtaking. Even these massive layers, some several tens of centimeters thick in places, were sulfur rich and loaded with blueberries. But there was a major component of volcanically derived basaltic particles, too small to resolve with the Microscopic Imager, in these layers as well. The way the layers conformed to one another, and the way groups of layers intersected other groups of layers at sharp, high angles, was reminiscent of similar layering seen on Earth in places where ancient sand dunes had been buried and compressed. Some of the alignment of the layers was probably related to the enormous tectonic forces associated with the impact that created Endurance Crater. It was hard to untangle these effects.

There were critical places that we wanted to reach to make the measurements necessary to distinguish among competing ideas, but we couldn't get the rover to many of these places because of the slopes or the sand. Ultimately, we were left thinking that what we were seeing may be a mix of ancient basaltic sand dunes and dunes composed of sand and eroded outcrop rock. We know that the outcrop rock is soft and has been eroded away over time because the blueberries have piled up on the surface. What we don't know, however, is where all the eroded outcrop "dust" and outcrop "sand" have gone. Perhaps they formed sulfur-rich sand dunes long ago that covered large parts of this region. All that's left of them today, though, may be their compressed and buried remains, exposed in special places like the walls at Burns Cliff.

Having gone about as deep and far around inside Endurance Crater as we could, we turned *Opportunity* around and started looking for a way out. We found one, not far from where we came in and performed another set of measurements on a different subset of the same layered rocks. Climbing up the slopes of Endurance was just as easy for the rover as climbing down had been. Before we knew it we found ourselves back out on the flat, seemingly endless plains of Meridiani. We had spent more than six months inside Endurance Crater. We had collected enough information to know that the watery past history of Meridiani Planum was extensive, in space (at least between Eagle and Endurance) and in time (at least as deep into Endurance Crater as we could get). Still, on the larger scale of things, we really hadn't sampled that much of Meridiani. We'd only driven across less than a kilometer of terrain, and we really hadn't gone that deep into the sedimentary record of the region (only sampling about ten meters' deep into Endurance). Now we would have to venture farther into the barren, desolate plains.

FOLLOWING SPREAD

160–161
Opportunity's second color panorama of Endurance Crater, revealing the steep cliffs along some of the walls. Pancam images acquired on SOLS 117 to 123. (NASA/JPL/CORNELL)

162–163
Oppurtunity's "Burns Cliff" panorama. This postcard spans 180° along one of the steep cliffs inside Endurance Crater, and was shot by Pancam on SOLS 287–294. (NASA/JPL/CORNELL)

164–165
False-color (for mineral analysis) postcard view of the 50-meter-wide sand dune field at the bottom of Endurance Crater, photographed by Pancam on SOL 211. Some of the dunes here are more than a meter tall from crest to trough. (NASA/JPL/CORNELL)

Mars Time

We came out of Endurance near the southern end of the crater, just before the 2004 end-of-year holidays on Earth. The team was tired from the constant demands of operating two spacecraft that had survived nearly four times their expected lifetimes. We knew we had some of the most incredible jobs in the world. But the rovers don't care about nights or weekends on Earth. They don't care about holidays, or your kids' birthdays, or your anniversary. They don't care if your son had to have an emergency appendectomy, or that you had to deal with weeks spent away from your family every month. All they care about is waking up, getting their commands, measuring, photographing, and radioing the results back to Earth, and then going back to sleep. Relentlessly. Like robots!

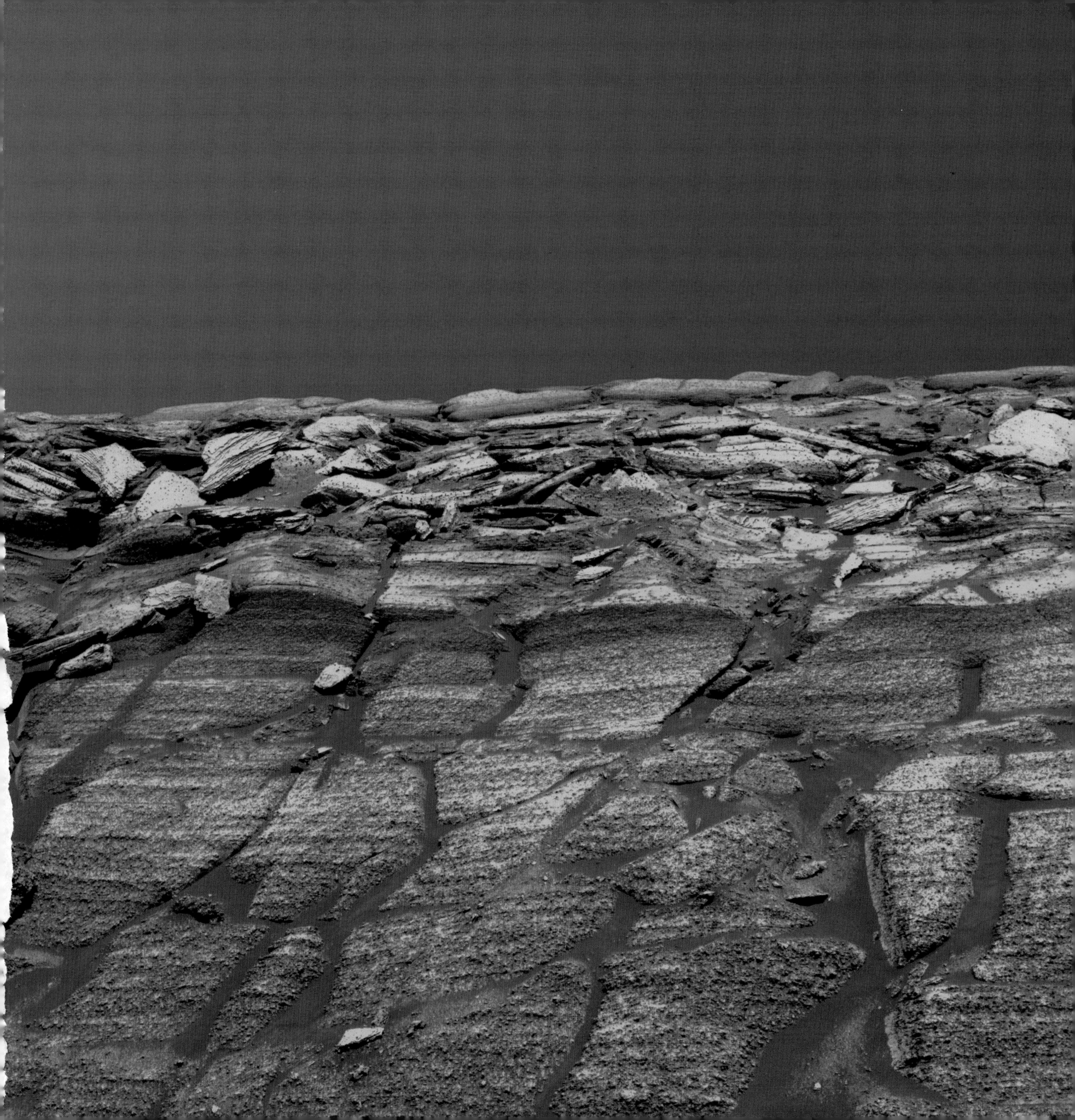

When we designed the plan for mission operations back in 2002 and 2003, we had assumed that roughly one day out of every three would be a "blown sol" where something would go wrong (a communications error, some unforeseen driving obstacle, some human commanding error, etc.). This estimate was based on previous missions' operations experience. The reality is that blown sols have been extremely rare, averaging something more like one in fifty instead of one in three. This is a testament, of course, to the robust engineering, software, and operations design of the rovers, to the quality of the mission operations staff at JPL, and to the hard work of the instrument teams at universities and laboratories around the world. But it's made for a grueling run.

JPL Project Manager Pete Theisinger, one of the many heroes who made this all possible, never failed to remind us that a ninety-sol mission worked out to about $4 million per rover per SOL to be on Mars. At the end of every sol, as the Sun set in the west and each rover shut down to go to sleep, we'd start poring over the data back on Earth, frantically studying the pictures and figuring out what to do with the rover tomorrow. We all had to look ourselves in the mirror and ask: "Did everyone get their $4 million worth of science today?" That number went below $1 million per SOL once we'd

survived more than four times our expected lifetime, but still...We were driven, and there were no excuses to slow down.

Nevertheless, by the end of 2004, the schedule had actually calmed down significantly relative to when we landed. From January through April that year, most of us had been living on "Mars time." Mars time is a strange schedule that's synced to sunrise and sunset on Mars, not on the Earth. It turns out that the time from sunrise to sunrise on Mars, a sol, is about forty minutes longer than an Earth day. Say the rover wakes up just after sunrise on one SOL and it happens to be 12:00 noon in Los Angeles. We radio that day's bundle of commands to the rover; it goes off and executes them, and radios the results back to Earth just before sunset, say around 9:00 p.m. in L.A. Then the next sol, the rover will wake up forty minutes later, at 12:40 p.m. in L.A. We'll get that day's images and other data back on Earth at 9:40 p.m. Then 1:20 p.m./10:00 p.m. the next day, then 2:00 p.m./10:40 p.m., and so on, until a few weeks later the rover is waking up in what is the middle of the night in L.A. Slowly, the schedule creeps back into the normal Earth workday schedule, then passes it, and then we're working third shift again. Being on Mars time was a very bizarre way to live and work.

LEFT TO RIGHT
Opportunity's "footstep" in crusty soil. Pancam SOL 605 photo. (NASA/JPL/CORNELL)

False-color (for mineral analysis) *Opportunity* Pancam SOL 496 photo of the steep, 20-cm-deep walls of Purgatory trench. (NASA/JPL/CORNELL)

LEFT TO RIGHT
Pancam postcard of layered, patterned outcrop rocks near the rim of Erebus Crater. *Opportunity* SOL 582. (NASA/JPL/CORNELL)

Close-up of "Payson," one of the few exposed parts of the rim of Erebus Crater. *Opportunity* Pancam SOL 657 postcard. (NASA/JPL/CORNELL)

It was often difficult to deal with Mars time in a pragmatic way. Everyone had to buy hardware or use software for some kind of Mars time clock or watch, for example. It was difficult in a physiological way as well. As humans who grew up on this planet, we're tuned in to the circadian rhythms of sunrise and sunset. Our bodies and brains have a "pulse" that works on an Earth time clock. Mess with that, even by only forty minutes per day, and sleep suffers, performance drops, and people make mistakes.

But then there are two rovers, separated by twelve Mars time zones on opposite sides of the planet. One wakes up as the other goes to sleep. Some of us were essentially working all the time, only getting little snippets of sleep during the morning or evening twilight hours for each rover before the Sun got high enough in the sky to wake one up or low enough in the sky to shut the other down. Methods of dealing with this included coffee and other tactical uses of caffeine, power naps, writing rover poetry, going away to sleep for days on end, choosing special "rover wakeup" songs for the start of each day, or eating way too much of the free ice cream that JPL was providing to us. And then there were pranks. One morning the Mini-TES team showed up in their workroom and all the keys on their computer keyboards had been mysteriously removed except the ones that could be used to spell "pancam." Another day the Pancam team showed up to work and the floor of our workroom was covered with cups of water, filled to the meniscus, with "MTES" spelled out in the middle of the room. The Mini-TES team showed up for work another day and all of their chairs, keyboards, monitors, and phones had been shrink-wrapped. The ice-cream freezer would mysteriously move from the fourth to the fifth floor, and then back. The red office chairs used on the *Spirit* floor were swapped

with the blue office chairs that were supposed to be on the *Opportunity* floor. It was all harmless fun, and it helped inject some levity into what was often a stressful experience.

Living on Mars time was an interesting and unique experiment. It made sense in terms of operating robots that were actually on the planet Mars, but despite the occasional fun, it was killing us. Caffeine, adrenaline, and chocolate ice cream can only carry you so far. This was partly why in May 2004, the project management switched us to a hybrid Earth-Mars time. We worked a fixed schedule between about 7:00 a.m. and 7:00 p.m., Pasadena time, for both rovers. We occasionally planned several days in a row, or operated in some kind of restricted fashion, during those times when we should normally be working in the middle of the night. This switch made a huge psychological as well as physiological difference on the team. It was like returning to Earth after a long trip away. Families were reunited, some measure of predictability came back to people's lives, and occasionally we could even sleep.

Everyone got a real break in September 2004 during the ten days or so when Mars was passing too close to the Sun, opposite the Earth on the other side of the solar system, for radio signals to reliably be transmitted between the planets. The schedule was further relaxed in October 2004 when we switched to a "weekends off" staffing system, spending some extra time on Friday to build rover plans for three sols on Mars (the equivalent of Saturday, Sunday, and Monday on Mars for each rover). This gave everyone a little bit more of their lives back, except for a small number of hardy and not-thanked-enough people at JPL and at NASA's Deep Space Network antennas in California, Aus-

tralia, and Spain who occasionally have to stay on Mars time because they need to communicate directly with the rovers.

Among the Wreckage

Opportunity spent most of the 2004 holiday season poking around among its own wreckage. We drove south from Endurance to check out the site where the heat shield had crashed when we landed back in January. We were interested in the site scientifically because we knew that the freshest impact crater on Mars would be there. The landing system engineers were also interested in seeing what the exterior of the heat shield looked like—how hot it had gotten, how much material had burned off the outside. They knew it had done its job, but that maybe they could make future heat shields do their job even better by examining this one up close.

The crash site was about 200 meters south of Endurance. It took about a week to drive the rover there, and when we arrived, the scene was surreal. Metallic parts were all over the place: springs, bolts, brackets, shards of insulation and aluminum thermal blanketing. Nearby we found a big splash mark on the ground outlined in bright red dust where the heat shield had actually hit. It's a strange crater: about 2.5 meters wide but only about 5 to 10 centimeters deep. No one's been able to figure out exactly how the heat shield, a hollow half-cone-shaped hunk of metal, was able to make such an oddball crater. We were able to avoid driving over most of the debris and obtained some excellent close-up images of the main remaining pieces, showing how the heat shield materials had burned and providing hopefully useful data for future designs.

It was a strange few weeks, poking around among our alien wreckage. We even found, to our amazement, a large iron meteorite right next to the main heat shield fragment. At least it looked like an iron meteorite. It was about the size of a loaf of bread. We weren't sure what it really was, though, until we put the Mössbauer spectrometer on it and saw a huge signature from nickel and iron pop up in the data. Yup, an iron meteorite. Sitting out here in the plains. What were the odds? Someone on the team commented that we should leave this place soon, because this was obviously the part of the planet where large metal things fall from the sky.

There was one other interesting aspect to the heat shield campaign, and it is an example of how controversial public policy issues like national security and the war on terror can directly influence government-run operations, even Mars rover missions. Before we got to the heat shield there was some discussion of not allowing the images of the heat shield to be released to the public or to the science team. Although no real details were given to us, some people were apparently worried that some aspects of the heat shield design might provide information about parts of classified weapons systems or other devices with similar designs and with national security implications. Federal laws prevent such information from being shared with most foreign nationals. The prospect of taking images on Mars that we couldn't show to anyone except people with classified access was extremely distressing to me. The likelihood that anything seen in an image of wrecked spacecraft parts on Mars could be used by someone to do harm to the free world seemed remote at best. We had been sharing all of our images with anyone with Internet access from sol 1. Fortunately, sanity won out and the issue did not come to a debate. We could release all our images just as we always had done.

LEFT TO RIGHT
Pancam blue-filter photo of layered outcrop crops viewed in late-afternoon sunlight on the rim of Erebus Crater. *Opportunity* SOL 665. (NASA/JPL/CORNELL)

Finely layered outcrop rocks photographed by Pancam at low Sun on SOL 690. Curved, festooned patterns in these images provided the best photographic evidence discovered thus far for liquid water along *Opportunity*'s traverse in Meridiani Planum. (NASA/JPL/CORNELL)

FOLLOWING SPREADS
172–173
Vertical projection view, simulating looking down on *Opportunity* out to a distance of 25 meters. False-color (for mineral analysis) Pancam postcard from images shot on SOLS 652–666. (NASA/JPL/CORNELL)

174–175
A Pancam triptych from "Olympia," northwest of Erebus Crater. *Opportunity* SOL 634. (NASA/JPL/CORNELL)

Back on the Road

Our goal after encountering the heat shield was to drive *Opportunity* south. We wanted to explore the boundary between the dark, blueberry-filled plains that we'd been on so far and what we could tell from orbital images was a brighter, perhaps "etched" kind of terrain that is also common in many parts of Meridiani. We wanted to characterize both of these kinds of terrains to give us a better picture of what this part of Mars is like overall. A secondary objective, if we were lucky to survive that long, was to eventually get to an even larger crater, called Victoria, about 6 km to the south. We hoped to try to sample even deeper layers of Meridiani's sedimentary rocks within that crater. Just as we had done for the drive from Eagle to Endurance, we mapped out a path using small craters as waypoints on the lonely sea heading south. This path would take us past more craters named after famous ships of exploration: Argo, Alvin, Jason, Vostok, James Caird, Viking, and Voyager.

This was when we put the pedal to the metal with *Opportunity*. The distance between the heat shield and the poorly defined boundary between the darker and the brighter terrains was about 4 kilometers. It was mostly flat, beautiful nothingness, punctuated every so often by a small crater, fissure, or sand dune. The goal was to get through that part of the traverse as quickly as possible. We spent nearly every SOL driving for most of the time that the rover was awake. On SOL 410 (March 20, 2005), we set the Mars single-sol driving distance record of 220 meters. Occasionally we would do some relatively simple surface and atmospheric monitoring or stop to examine the chemistry and mineralogy of random pieces of exposed rocks. We were seeing the same basic sulfur-rich, blueberry-laden sedimentary rocks, buried about a meter (or less) beneath a concentrated layer of blueberries, sand, and small basaltic rock fragments. The surface became a pattern of low ripples of sand resembling waves on the ocean. As we headed farther south, the average height of the ripples started to increase, from about 5 to 10 cm high near the heat shield to 10 to 20 cm high by about 3 km to the south. Curiously, it seemed like the blueberries were actually getting slightly smaller, on

LEFT TO RIGHT
A postcard of the Eagle Crater outcrop, photographed by the *Opportunity* Pancam on SOL 3.
(NASA/JPL/CORNELL)

Opportunity's SOL 1 postcard —Pancam's first color images from Meridiani Planum.
(NASA/JPL/CORNELL)

average, as we headed south. Our measurements were pretty sparse, though, so it was hard to tell.

Stuck in Purgatory

Everything was going beautifully and we could almost start to smell the bright terrain just a few hundred meters ahead. All hell broke loose, however, when the SOL 446 (April 26, 2005) photographs came in. We had been driving *Opportunity* almost due south, crossing over hundreds of small sand ripples for nearly three months. Suddenly, we found ourselves stuck, buried "up to the hubcaps" in fine, dark sand. It had been a long time since we'd heard audible gasps among the team when the images first came in. It was kind of like the old days, and not in a good way. The rover had apparently attempted to drive across a particularly deep sand ripple, perhaps 25 to 30 cm tall, but it wasn't able to get beyond the crest. *Opportunity* was bogged. The rovers have automatic sensors that will stop a drive if the wheels start running into too much resistance or exerting too much torque, or if the suspension system starts tilting outside of some predefined limits. In this case, though, the sand was so fine and so slippery that the wheels just kept spinning and spinning. From the rover's perspective, everything was fine—the wheels weren't meeting any particularly strong resistance and it appeared to be staying roughly level. So it kept "driving" along, sinking and sinking, digging deeper and deeper ruts into the sand. The rover had been commanded to drive about 90 meters that day; it drove about 40 meters, encountered the

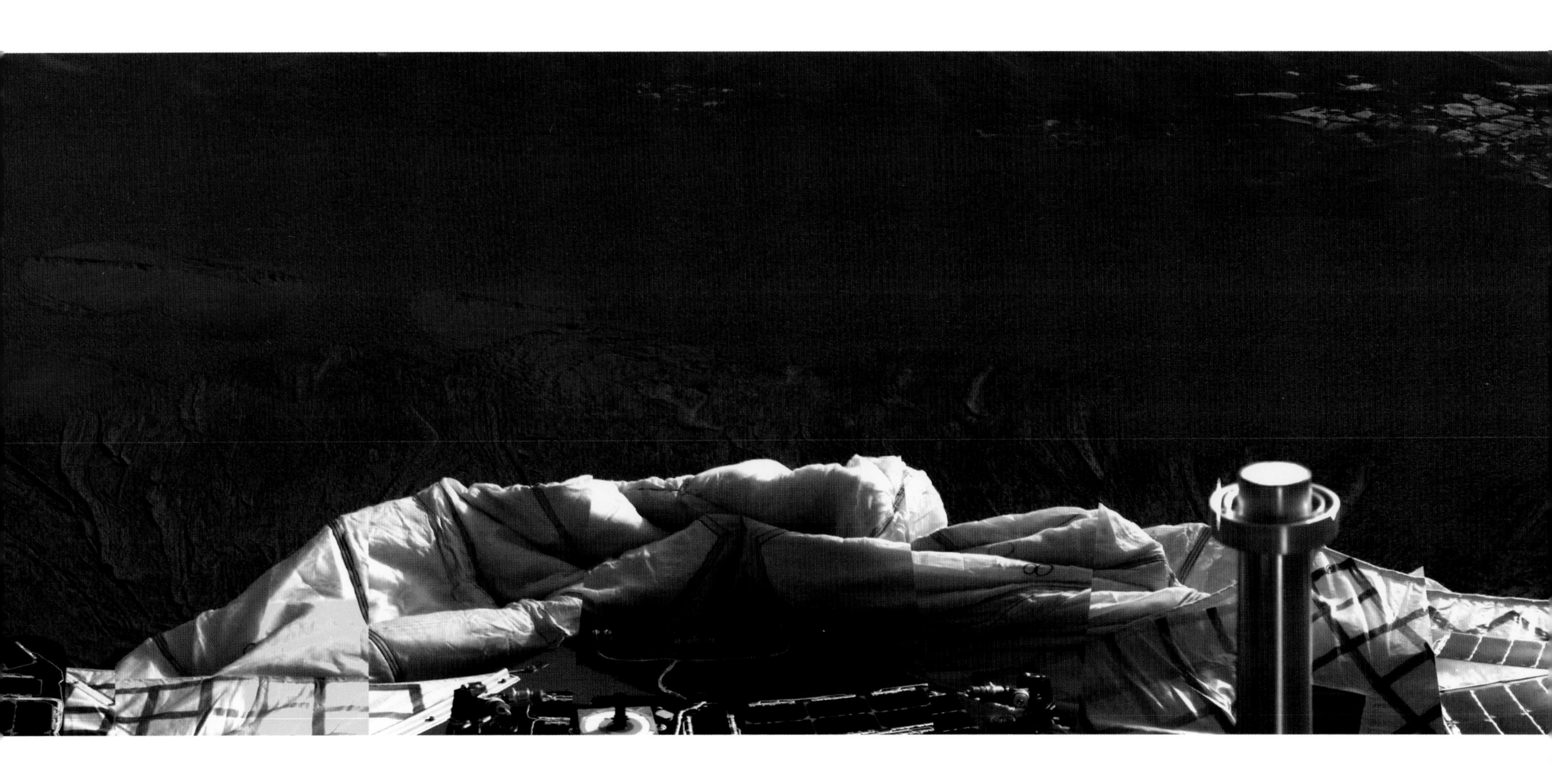

deep sand, and then sat there spinning its wheels for the equivalent of another 50 meters.

The rover's instruments and cameras were all healthy. The engineering and mobility teams sprang into crisis mode while we used the driving respite to photograph an enormous color Pancam panorama of this part of the plains. We named it the "Rub al Khali" pan, after the Arabic words for the "Empty Quarter" region of Saudi Arabia, which seemed fitting given our surroundings and predicament.

The engineers took the test rover at JPL and buried it in fine sand the same way *Opportunity* was buried, using a combination of sand, clay, and sticky diatomaceous earth to try to simulate the fine-grained nature of the Meridiani ripples. They tried various drive strategies for escaping. The scientists also got heavily involved in the escape planning, led by Cornell colleague and team member Rob Sullivan, an expert in sand dunes, ripples, and wind-related or "aeolian" processes on Earth and Mars. Rob and the engineers tried various combinations of wheel wiggles, rocking motions, attempted turns, and driving at various speeds (in the typical 1 to 5 cm/sec "speed" range). Time and again they found that the best strategy was an age-old one: Slow and Steady Wins the Race.

The key to escaping from the sand, it turned out, was to command a limited drive distance each SOL and accept the fact that the rover would slip more than 99 percent of that distance, but would make *some* progress. For example, the drivers would command a 20-meter drive backward out of the rut that we had created and expect the rover to move only about 10 cm. This would be good

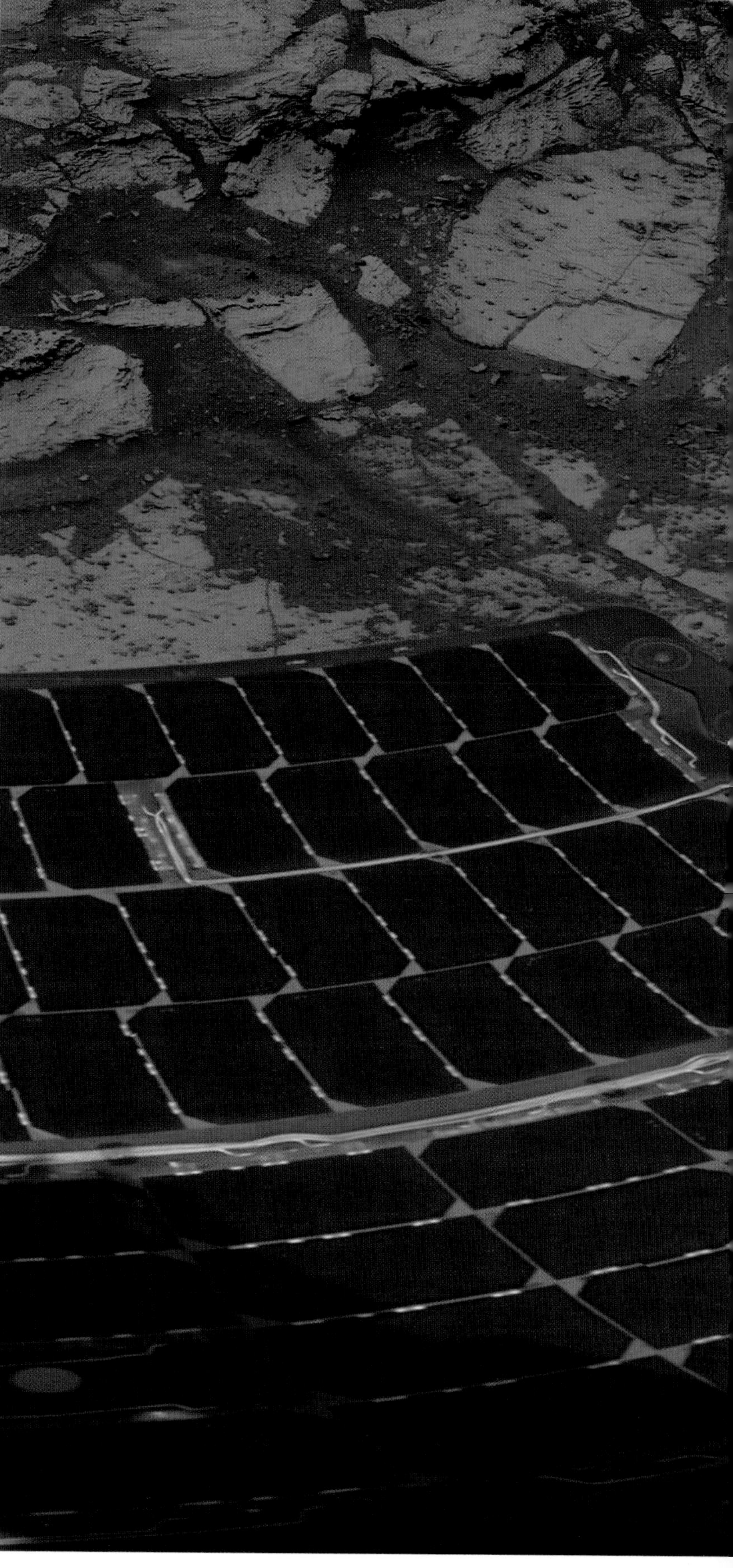

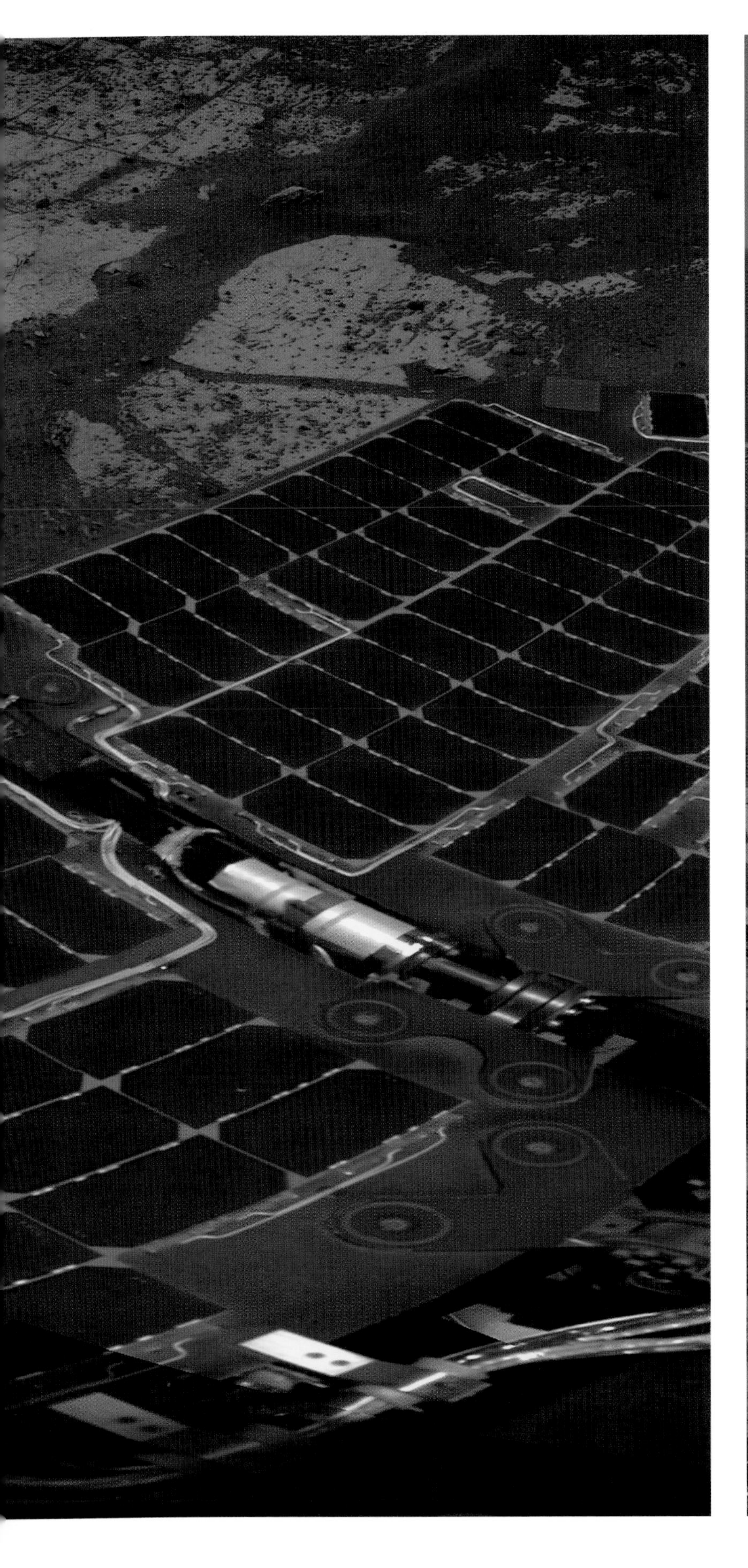

progress. The theory, based on work with the test rover on Earth, was that eventually the wheels and cleats would gain enough traction and we would come zooming out of the rut. People who've ever gotten their car stuck in the snow know this routine well. The test rover came zooming out of the rut every time the drives were done this way, eventually; it just wasn't predictable whether it would take a few days or a few weeks.

Opportunity was commanded to creep along, backward, slipping more than 99 percent every sol, for nearly 40 sols. We started calling the place "Purgatory Dune" because we didn't know if we would ever really escape. Some of the scientists, engineers, and managers on the team, and even some people at NASA Headquarters, were getting nervous. Two weeks passed, then a month. Would *Opportunity* die here? But Rob and the other mobility experts continued to counsel patience: Slow and steady...slow and steady...finally, on SOL 484 (June 4, 2005), the rover "zoomed": scooting backward out of the rut about a meter after 14 meters of commanded driving. We were free! It was a good feeling to be mobile again. Even the optimists on the team were starting to get antsy about possibly becoming a "lander" mission in the middle of such a desolate spot.

The critical lesson of getting stuck, though, was that we hadn't seen it coming. Sure, the ripple that we got stuck in was a bit taller than average, but otherwise there was nothing particularly unusual about it or its surroundings. After getting unstuck, we tried to find some distinguishing characteristic about Purgatory Dune, something like unusual color, or texture, or composition, but we couldn't find anything especially different. Maybe it was just bad luck, or maybe there was just something about this transitional terrain that was different but that we just didn't have the ability to measure or the brains to figure out. Whatever the reason, driving after Purgatory became a much slower, more careful, methodical activity. The rover drivers would try to drive in the troughs between ripples and then, cautiously and gingerly, cross over a ripple crest only when absolutely necessary. Gone were the 200-meter drive days.

PGS 178–179
Triptych of the *Opportunity* "Erebus Rim" 360° panorama, including the rover deck. This was the largest Pancam panorama acquired by either rover. It was shot over 43 SOLS (652–694) while problems with the rover arm were being diagnosed. This mega-postcard consists of 1,590 images shot through fourteen of Pancam's filters. (NASA/JPL/CORNELL)

FOLLOWING PAGE
A stark landscape of outcrop and sand just north of Erebus Crater. *Opportunity* Pancam SOL 593 postcard. (NASA/JPL/CORNELL)

Rolling On

Opportunity continued to make progress to the south, eventually reaching the rim of a crater called Erebus that is about double the size of Endurance Crater. Erebus is wide but shallow, and we discovered it to be almost entirely filled by sand. Along the rim of Erebus, however, is a marvelous "highway" of flat, layered, reddish, sulfur-rich outcrop rocks. These rocks are very similar to the ones that we saw at Eagle Crater and Endurance Crater, with one important difference: In many of them, the blueberries are much smaller or even gone! We don't understand yet why the berries are disappearing as we head south, but it's high on our list of things to figure out.

When *Opportunity* reached the rim of Erebus Crater in the fall of 2005 we encountered another problem. After a series of science measurements one day, the rover's arm failed to stow properly. A problem was discovered with the motor that moves *Opportunity*'s shoulder joint—the same motor, in fact, that is near the heater that has been stuck on for most of the mission. The engineering team sprang back into action to diagnose the problem and develop work-around solutions. While they did, we spent the time acquiring a huge new 360° Pancam panorama through all of the camera's filters (the first time we've been able to do that on either rover), as well as acquiring late-afternoon low Sun images of spectacular new festoons and other sedimentary structures discovered in the layered rocks near Erebus. Ironically, our forced stop has revealed some of the strongest evidence yet confirming the presence of shallow liquid water on the surface of Mars in this region long ago. Once the arm problem diagnosis and repair/work-around is

complete, we will continue our photographic reconnaissance of Erebus Crater, then get back on the highway heading south toward our next major goal: Victoria.

As we continue to explore new vistas in this special part of Mars, I find myself wondering what lies in store next for our intrepid little robot explorer. That lazy, bad-shot "sniper" is still out there. Someday maybe he'll get us, but there's still no obvious end of the mission in sight. Despite the occasional aches and pains, nothing on *Opportunity* (or *Spirit*, for that matter) is slowly degrading or "running out," although many of the motors and other devices have exceeded their design lifetimes. There's no way to know when the adventure will end. The solar panels continue to provide high levels of power for driving and science observations because Meridiani is a much less dusty place than Gusev, and so *Opportunity* is a much less dusty rover than *Spirit*. So we press on. Victoria is nearly 800 meters (a half mile) wide and perhaps 50 to 100 meters deep in places. We're hoping that even more of the subsurface will be exposed for view. Will we find the limits of Meridiani's watery past? Will we find out how long the water was there, or what happened to it?

Opportunity's exploration mission in Meridiani Planum began when planetary scientists discovered a unique mineral signature there in measurements made from one of the NASA Mars orbiters. It has culminated, so far, in the discovery that there was liquid water involved in the formation of those minerals, and of others, on this part of the planet long ago. That water was on the surface, or very near the surface, for some unknown but geologically significant period of time. The basic forensic evidence has been and continues to be collected. Now the game is trying to figure out how to reconstruct the scene of the crime.

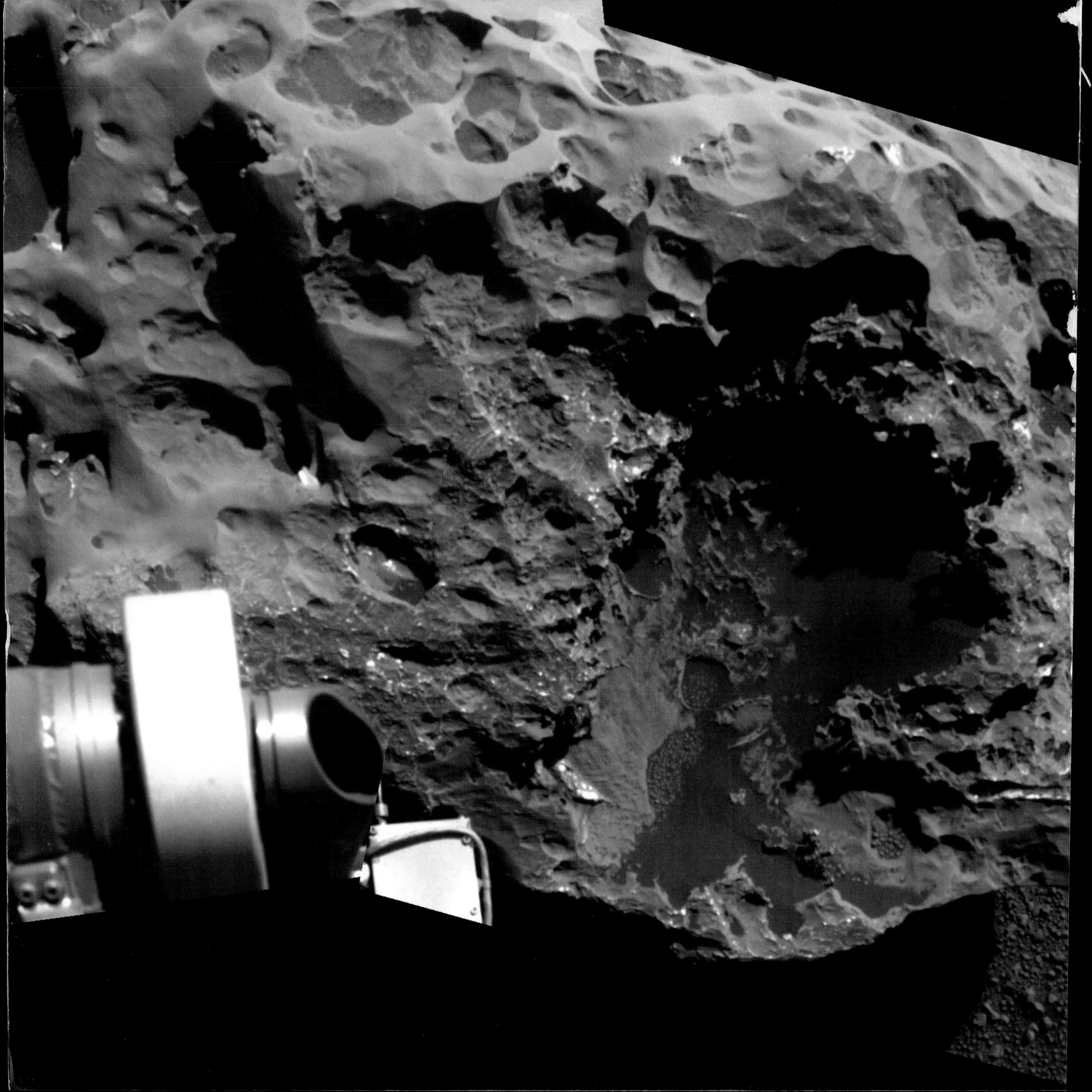

Postscript

For more than six years now the postcards have just kept coming in from *Spirit* and *Opportunity.* The rovers have far, far surpassed our every expectation, now having sent back more than 250,000 photos of the surface of Mars, taken millions of infrared spectra, made more than 1,500 detailed chemical and mineral measurements with the arm instruments, and driven more than 27 km (17 miles) combined! Thus, it seemed fitting to update this paperback version of the book to include some of the most dramatic, picturesque, and scientifically compelling photos sent back from the rovers since the hardcover edition came out in 2006. These new views of ever more astonishing alien vistas are the book's real postscript.

Both rovers continue to operate well—though with challenges—and both have made important new scientific discoveries. *Spirit* has spent the last four and a half years exploring the region on and around Home Plate, acquiring photos and other data that have led to the remarkable discovery that the landscape may be the eroded remains of an ancient hydrothermal system. The evidence comes partly from the large number of dark, vesicular (glassy, frothy) volcanic rocks found all around Home Plate, and partly from the fine-scale layers of soft volcanic ash that have been mapped around the edges of the feature. The clincher, though, has come from a classic case of victory snatched from the jaws of defeat: While approaching Home Plate on SOL 779, *Spirit*'s right front wheel motor failed, rendering the wheel unable to spin. We were forced to have to drive the rover backward, dragging that wheel through the soil like a gimpy leg, and severely limiting our mobility to only 5 to 10 meters per day. However, the dragging wheel excavated a long trench as we drove and—to our great surprise—it began digging up bright white and yellow soils from the subsurface. Chemical and mineral analysis revealed the soils to be rich in hydrated sulfate minerals and amorphous, hydrated silica. Discovery of these kinds of minerals, combined with the volcanic rocks and ash layers, provide compelling evidence that Home Plate was once an active hydrothermal environment, that is, a hot spring, like the geysers in Wyoming's Yellowstone National Park. Such environments on Earth offer perfect niches for a variety of bacterial and more complex life forms. While we don't know if the region around Home Plate was ever inhabited, the evidence that it was at least *habitable* to life as we know it has been one of the most important discoveries of *Spirit*'s long, successful mission. But if the wheel had not broken, we may never have found those telltale mineral deposits.

As *Spirit*'s fourth Martian winter began to set in, the rover encountered its most challenging

LEFT
Opportunity continues to encounter iron and nickel meteorites while driving across the plains of Meridiani. This Pancam SOL 1961 false-color (for mineral analysis) mosaic shows a close up of one called Block Island, which is about 70 cm (28 in) across. (NASA/JPL/CORNELL)

predicament yet. While driving along the west edge of Home Plate we unknowingly drove onto some crusty soils that were covering a small impact crater. The weight of the rover broke through the crust and four of the wheels sank into the soft, salty sand below. We were stuck—just like *Opportunity* had been at Purgatory. Several months of driving efforts weren't able to get us unstuck, but the team was able to make some progress using clever methods of combining back-and-forth wheel wiggling and rolling. While we were starting to make progress getting out, the low winter Sun and dusty solar panels dropped the power levels so low that more drive attempts were no longer possible. Instead, we're just going to have to hunker down and concentrate on keeping *Spirit* alive during this most challenging of winters, in the hopes that the returning springtime Sun will once again allow us to drive and extract ourselves from our sandy trap.

Meanwhile, *Opportunity* has continued to roll, roll, roll across the plains of Meridiani. Around SOL 952 the rover finally reached Victoria Crater, after nearly two Earth years of driving to get there. Imagine a two-year road trip—with no stops for gas. At Victoria we were treated to some of the most dramatic and beautiful landscapes that we have ever seen on Mars. Sheer cliffs, 10 to 25 meters high in places, revealed coarsely banded to finely layered rocks in a geologic panoply reminiscent of the most spectacular layered cliffs and canyons on Earth. We were able to drive a little ways down into the crater, far enough to tell that the sulfur-, chlorine-, and bromine-rich outcrop rocks, loaded with blueberries, continue to be found even more than 15 km from where we first found them in Eagle Crater, back in 2004. The alteration of ancient sandstones by surface water and groundwater was apparently a truly widespread, regional process in this part of Mars. Opportunity is now heading toward an even bigger goal: the 22-km-wide crater Endeavour, located about 12 km southeast of Victoria. We've been driving toward Endeavour for a year, and we're about a third of the way there. Along the way we've taken some exciting detours to study a field of iron-nickel meteorites discovered perched atop the sandy dunes and ripples of the plains. Another perched rock that we thought might be a meteorite turned out to be a piece of Mars itself, a piece of impact debris with a kind of volcanic composition that we had not yet seen from either rover, tossed there from some unknown crater elsewhere on the planet. While we will struggle to keep making forward progress toward Endeavour during the Martian winter, it could still take another 12 to 18 months to get there. A prize may await us: Orbital measurements show that there are deposits of clay minerals exposed along the crater's northwest rim, right along our drive path. Clays are minerals formed in watery environments, but unlike sulfates they are formed in much more biologically friendly, nonacidic waters. We've seen lots of sulfate deposits now, from both rovers, but we've never seen clay minerals up close. The kinds of clays waiting for us at Endeavour Crater are seen elsewhere on the planet, too, in other past watery environments, including the candidate landing sites for NASA's next rover, the *Mars Science Laboratory* (due to launch in 2011 and land in 2012). The promise of delicious new water evidence ahead has us all excited again—Endeavour or bust!

The rovers have vastly exceeded their design lifetimes, and no one knows when their Martian adventure will end. When kids ask me, "When will the rovers die?" I turn the question around and ask them, "When will your toaster die?" After a momentary quizzical look, the light flashes on in their heads: they are machines, and we don't really know when they'll stop working. One day, though, they will have each sent back their last postcard—a final farewell note from our seasoned wandering explorers, written in the bluish setting Sun of their eternal Red Planet resting place. In the meantime, they'll keep trucking on, and we'll keep experiencing the magic of Mars through their robotic eyes.

Spirit Pancam false-color (for mineral analysis) SOL 2114 postcard of a potentially volcanic hill called Von Braun, which could be the rover's next exploration target after completing the study of Home Plate. (NASA/JPL/CORNELL)

FOLLOWING SPREAD
186–187
Spirit Pancam false-color (for mineral analysis) panorama of layered rocks along the edge of Home Plate, photographed on SOLS 748–751 (NASA/JPL/CORNELL)

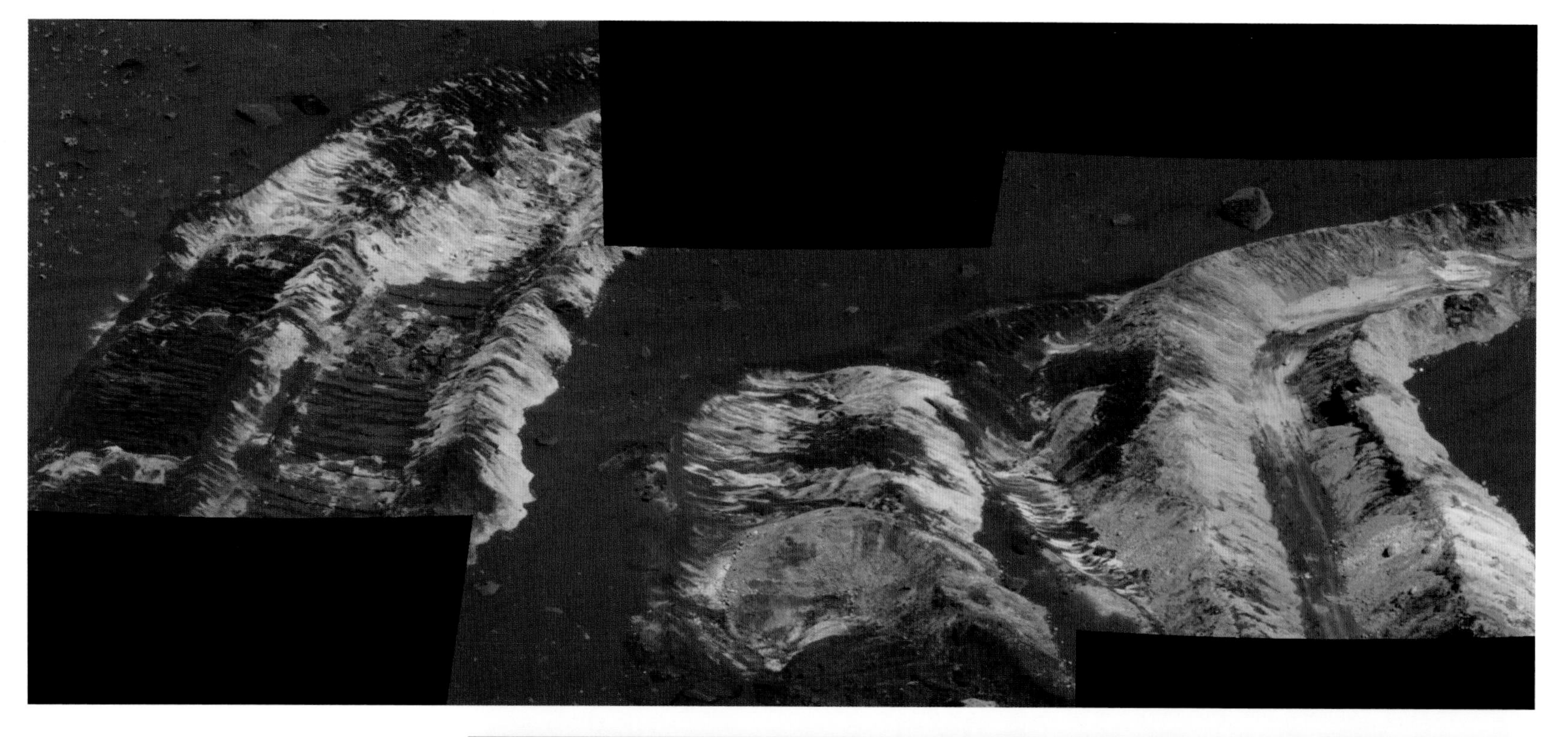

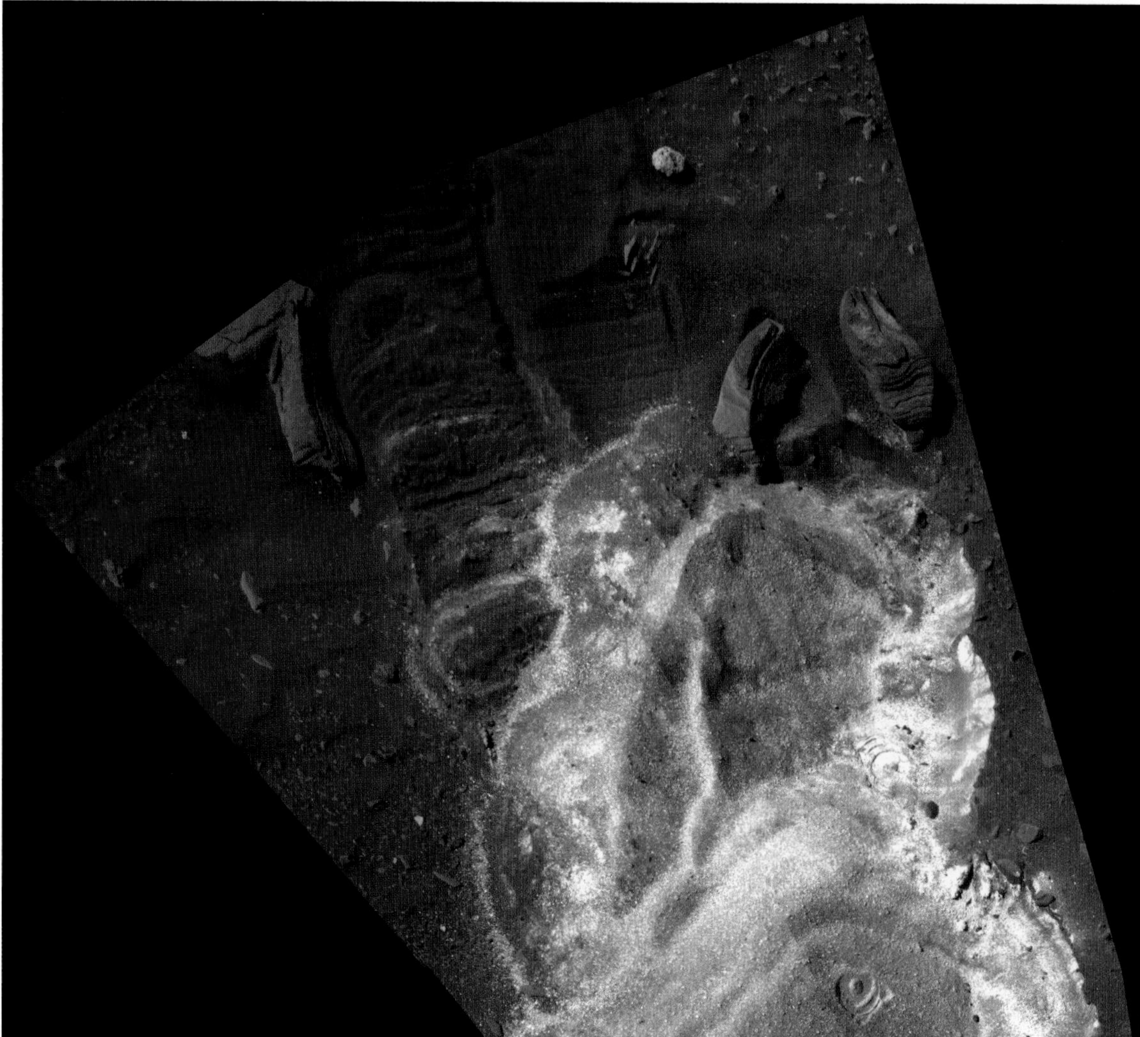

TOP TO BOTTOM
This *Spirit* Pancam SOL 788 mosaic shows a dramatic example of the whitish and yellowish salty soils dug up by the rover's wheels at a place called Tyrone, near Home Plate. (NASA/JPL/CORNELL)

In late 2008, *Spirit* got bogged down in the soft, salty soils around Home Plate. This Pancam SOL 1933 false-color (for mineral analysis) mosaic reveals the lovely variety of colors and textures in the soils where the rover is stuck. Several imprints can be seen where the rover has made detailed chemical and mineral measurements. (NASA/JPL/CORNELL)

FACING PAGE
Having completed our study of Victoria Crater, *Opportunity* is now heading toward an even larger crater called Endeavour. Even using new driving methods like the obstacle-avoidance software shown tested here in this SOL 1162 Pancam postcard, the rover might not get to Endeavour until sometime in 2011 or 2012. (NASA/JPL/CORNELL)

Acknowledgments

There are so many people to acknowledge and thank on so many levels for making these photographs, stories, and scientific results possible. Among my colleagues, I want to first acknowledge my friend and rover science team leader Steve Squyres, the Zeus from whom this particular incarnation of *Athena* really sprang, though certainly not fully formed. Among my most cherished memories of this project is that of sitting in the same room with Steve while we took a congratulatory phone call from the president of the United States (how often do you exchange *those* kinds of looks with your friends and colleagues?), and then a few days later, watching him give a passionate Mars exploration lecture to the vice president of the United States, one-on-one, in the rover command center at JPL. Talk about getting the message to the people at the top.

I also owe a large amount of thanks to all of my colleagues on the MER science team, comrades in the trenches during tactical martian operations, with special acknowledgments to Mike Malin, John Grotzinger, Phil Christensen, Jeff Johnson, Ken Herkenhoff, Mark Lemmon, Ray Arvidson, Larry Soderblom, and Dick Morris for mentoring, reviewing, and generally keeping me honest about what I thought I knew about martian geology, geochemistry, remote sensing, and space flight instruments. I also am indebted to, and in awe of, the design and engineering teams, flight software team, and operations team at JPL and other universities and government labs who have made these missions happen. Thousands of people were involved in the rovers, and all of them should feel a special sense of pride in their accomplishments. Special thanks go to project and mission managers Pete Theisinger, Richard Cook, Jim Erickson, John Callas, Barry Goldstein, Jennifer Trosper, ATLO managers Matt Wallace and Raul Romero, and many others for believing in the science goals and the science team, and for believing in your engineering and operations teams just as deeply. Thanks also to the Kennedy Space Center launch team and the JPL Entry, Descent, and Landing team for getting us there, and to Glenn Reeves and the Flight Software and Systems teams for getting us a CPU *that did not crash*. (Okay, it did once, but you fixed it pretty quickly. . . .)

So many people worked incredibly hard to bring these photos to life. In no particular order, and with humble apologies to the people I may have missed, I want to acknowledge Dave Brown, Mark Schwochert, Dave Thiessen, Darryl Day, Andy Collins, Tom Elliot, Arsham Dingizian, Ali Pourangi, Pete Kobzef, Greg Lievense, Alejandro Soto, Nancy Cowarden, Joy Crisp, Calvin Grandy, Mary White, Bobbie Woo, Larry Scherr, Greg Smith, Larry Soderblom, Len Wayne, Art Thompson, Leo Bister, and the rest of the highly capable ATLO staff for their countless hours of assistance and advice regarding the rover camera design, testing, assembly, and calibration. I also thank Dan Britt, Woody Sullivan, Larry Stark, Jon Lomberg, Lou Friedman, Tyler Nordgren, Bill Nye, Dick Morris, Mike Shepard, Roger Clark, Patrick Pinet, Yves Daydou, and Pascal Depoix for their assistance with building, testing, and characterizing the Pancam calibration targets (MarsDials) and materials, and Lisa Gaddis, Jim Torson, Jeff Johnson, Larry Soderblom, Doug Ming, Bill Farrand, Mike Wolff, Scott McClennan, Ed Guinness, Frank Seelos, Cathy Weitz, John Grant, Morten Madsen, Walter Goetz, Roger Tanner, Zoe Learner, Jason Soderblom, Yong Shin, Salman Arif, and Stephanie Gil for helping to calibrate the cameras so that we could get the colors right. Also critical to the success of the Pancam imaging has been the group of dedicated and hardworking "Payload Downlink Leads" and "Payload Uplink Leads" (responsible for instrument health monitoring, tactical data product generation, and daily sequencing of new images). This group includes Jonathan Joseph, Jascha Sohl-Dickstein, Heather Arneson, Miles Johnson, Dmitry Savransky, Bill Farrand, Walter Goetz, Alex Hayes, Ken Herkenhoff, Kjartan Kinch, Morten Madsen,

FACING PAGE
The exploration of Victoria Crater by *Opportunity* revealed some of the most dramatic and picturesque landscapes yet encountered by either rover. This Pancam SOL 1167 false-color (for mineral analysis) mosaic shows some of the steep, layered cliffs of the promontory called Cape of Good Hope. (NASA/JPL/CORNELL)

Justin Maki, Elaina McCartney, Dick Morris, Tim Parker, Jon Proton, Frank Seelos, Jason Soderblom, Rob Sullivan, Emily Dean, Mike Wolff, Dale Thieling, Tim McCoy, Alian Wang, Melissa Rice, Ryan Anderson, Justin Hagerty, Matt Chojnacki, and Eldar Noe Dobrea. Jeff Johnson and Mark Lemmon were also part of this group, and they deserve special mention for often filling in for me as the Pancam group leader during development, testing, and operations on Mars when other personal or work duties called me away. An enormous amount of gratitude also goes out to the "Cornell Calibration Crew," a dedicated and hardworking team of undergraduate and graduate students and staff members who have been carefully monitoring camera performance and calibrating Pancam images by staring at MarsDials every sol since January 2004. Thank you to Graham Anderson, Shianne Beer, Diane Bollen, Lindsey Brock, Rich Chomko, Adam Fischman, Kristen Frazier, Alissa Friedman, Stephanie Gil, Lisa Grossman, Ben Herbert, Ben Holmes, Kelley Hess, Min Hubbard, Peter Meakin, Chase Million, Sarah Morrison, Mary Mulvanerton, Eldar Noe Dobrea, Lucy Ooi, Aaron Rubin, Diego Saenz, Alex Shapero, Alex Shih, J. R. Skok, Pam Smith, James Teague, Tom Ternquist, Christopher Versfelt, Janet Vertesi, Brandi Wilcox, Ashlee Wilkins, and Nicholas Wirth. I also want to thank the MIPL imaging gurus at JPL who kept the pixels flowing and helped stitch many mosaics: Bob Deen, Doug Alexander, Helen Mortensen, Payam Zamani, Oleg Pariser, Vadim Klochko, Jeffrey Hall, Amy Chen, and Kris Capraro.

I thank Joan Popolo, John Debruzzi, Karen Guerrette, Erik Fahlgren, and Joe Palca for helping to pave the trail of contacts that made this idea, then this book, come alive. I am also indebted to my agent Michael Bourret at Dystel & Goderich and my editor Stephen Morrow at Dutton for their constant enthusiasm, patience, and advice. What a fascinating, educational process it is to write a book!

Finally, I want to thank my family and our friends who helped us survive the ups and downs of this adventure. Maureen, Dustin, and Erin: Your love, understanding, and patience have helped keep me grounded despite spending far too much time away from home on airplanes, in laboratories, or on some cold, alien planet. Our neighbors and friends in upstate New York supported our family through medical emergencies and the many hassles of my long stints away from home. Thanks so much to Mark and Kris Armstrong, Anne and Carl Czymmek, Cathy and Eric Haines, Julia and Genna Samorodnitsky, Paula Swayze, Trish and Kyle Thomas, Cindy and Harold Van Es, and all your kids. To my extended West Coast family, especially Bob Thompson and Jan Fuehrer and Kris Ockert and their families, thanks for helping keep us sane and laughing as much as possible. Finally, to my extended family and friends taking up most of the state of Rhode Island, especially my parents, Jim and Angela Bell, my grandparents, and my sisters Wendy, Kristen, and their families, thank you for the love and support that set me on the path that led to all this. I still use that old telescope of mine, though not as much as I should.

Appendix

METRIC CONVERSION CHART		
1 kilometer = 0.62 miles	1 meter = 3.28 feet	1 centimeter = 0.39 inches
0°C = 32°F	-100°c = -148°F	

Additional Resources

MARS EXPLORATION ROVER WEB SITES

http://*Athena*.cornell.edu (details on the rover's instruments and mission)
http://marsrovers.jpl.nasa.gov/gallery/all (site where new rover images are posted)
http://marsrovers.jpl.nasa.gov (general mission information)
http://pancam.astro.cornell.edu (Pancam home page)
http://pds-geosciences.wustl.edu/missions/mer/geo_mer_datasets.htm (science data)

ROVER-RELATED MAGAZINE ARTICLES

Bell, J. F. "The Human Side of Mars Exploration." *The Planetary Report*, November-December 2003.
—— "Mineral Mysteries and Planetary Paradoxes." *Sky & Telescope*, December 2003.
Bell, Jim. "Blazing a New Path." *Astronomy*, August 2003.
—— "In Search of Martian Seas." *Sky & Telescope*, March 2005.
—— "Backyard Astronomy from Mars." *Sky & Telescope*, August, 2006.
—— "The Red Planet's Watery Past." *Scientific American*, December 2006.
—— "Portraits from Mars." *Astronomy*, January 2007.
Christensen, Philip R. "The Many Faces of Mars." *Scientific American*, July 2005.
Petit, Charles W. "Making a Splash on Mars." *National Geographic*, July 2005.

DETAILED SCIENTIFIC RESULTS FROM THE ROVER MISSIONS

Science magazine, August 6, 2004, vol. 305, no. 5685 (*Spirit* first results)
Science magazine, December 3, 2004, vol. 306, no. 5702 (*Opportunity* first results)
Nature magazine, July 7, 2005, vol. 436, no. 7047 (*Spirit & Opportunity* results)
Earth & Planetary Science Letters, vol. 240, no. 1, 2005
Journal of Geophysical Research, vol. 111, February, 2006

BOOKS, WEB SITES, AND OTHER GENERAL INFORMATION ABOUT MARS

Bell, Jim. *Mars 3-D*. New York: Sterling, 2008.
—— *The Martian Surface*. Cambridge: Cambridge University Press, 2008.
Carr, Michael. *The Surface of Mars*. New Haven: Yale University Press, 1981.
Croswell, Ken. *Magnificent Mars*. New York: Free Press, 2003.
Goursac, Olivier de. *Visions of Mars*. New York: Harry N. Abrams, 2005.
Jakosky, Bruce M. *The Search for Life on Other Planets*. New York: Cambridge University Press, 1998.
Kieffer, H. H., B. Jakosky, C. Snyder, and M. Matthews, eds. *Mars*. Tuscon: University of Arizona Press, 1992.
Squyres, Steve. *Roving Mars: Spirit, Opportunity, and the Exploration of the Red Planet*. New York: Hyperion, 2005.

http://nssdc.gsfc.nasa.gov/planetary/chronology_mars.html (Mars exploration)
http://nineplanets.org/mars.html (Mars facts)
http://www.marssociety.org (The Mars Society)
http://www.planetary.org (The Planetary Society)

Index

Note: Page numbers in *italics* indicate photographs.